Trees and Building Sites

Proceedings of an International Workshop
on Trees and Buildings

Edited by:
Dr. Gary W. Watson
The Morton Arboretum, Lisle, Illinois
and
Dr. Dan Neely
Editor, International Society of Arboriculture
Scott City, Missouri

Presented at the Morton Arboretum
Held May 31 through June 2, 1995

Published by
International Society of Arboriculture
P. O. Box GG
Savoy, Illinois 61804 USA

Printed in the United States of America

10 9 8 7 6 5 4 3 2 1

ISBN 1-881956-15-6

Where trade names, proprietary products or specific equipment are listed, no discrimination is intended, and no endorsement, guarantee or warranty is implied by the presenters and/or their affiliated organizations, the International Society of Arboriculture or The Morton Arboretum.

Cover design by the Morton Arboretum
Composition by Betty Ann Neely
Printed by Printec Press

Copies of this book may be ordered
directly from the publisher

The International Society of Arboriculture
P.O. Box GG
Savoy, Illinois 61874-9902 USA
Phone (217) 355-9411
Fax (217) 355-9516

Table of Contents

Part I — Tree Damage

Part II — Building and Infrastructure Damage

Part III — Tree Protection and Preservation

Introduction

In urban and suburban landscapes, trees almost always grow in close proximity to buildings and infractructure. The potential for conflict between trees and man-made structures is always great. The process of developing land for building structures and infrastructure can be devastating to existing trees. The changes in the environment resulting from developing woodlands can also have an impact on existing trees. Post-development soils are often so disturbed that new trees planted on the site have difficulty surviving. As trees do manage to establish and mature, they may eventually be viewed as responsible for damages to buildings and infrastructure. If sidewalks, sewer lines or foundations need repair because of trees, is it a tree problem, or is it a problem of poor management or tree-vulnerable construction? Finding solutions will require a multi-disciplinary approach.

The Trees and Buildings Conference was held, and these proceeding published, to establish a foundation of information for managing trees on building sites, upon which we can begin to build at an accelerated rate. Though the research base has expanded considerably in recent years, much more information is needed. Practitioners need scientific information from researchers, but researchers need feedback from practitioners on practical application of current research results and needs for additional research. A conference like *Trees & Buildings*, with presentations from research scientists and managers and an audience full of persons with rich experiences, provides a unique opportunity for the kind of international interaction that will stimulate progress in minimizing conflicts between trees and man-made structures. We hope that this was only the first Trees & Buildings Conference.

Acknowledgments

The success of Trees & Buildings Conference is due to the cooperation and support received from a host of organizations. A generous grant from the USDA Forest Service Urban Forestry Center for the Midwestern States made it possible to bring together such a large number of speakers, and to publicize the conference to planners, developers, builders and other professions outside of arboriculture. The cooperation between the Illinois Arborist Association and the Arboricultural Research And Education Academy (chapter and professional affiliation of ISA, respectively) in sponsoring the conference was significant. Professional affiliations are often in a position to provide technical expertise in a particular area, and chapters often have the structure required to organize a conference. The ISA and the Morton Arboretum also played important roles as supporting organizations.

PART I

Tree Damage

The Significance of Trees and Their Management in Built Environments

John F. Dwyer

The management of trees and forests in built environments is a major challenge given the complex interactions and high values involved, as well as considerable uncertainty over the outcome of management options. These challenges are outlined and improved approaches suggested.

Built environments include some of our most valuable trees and forests, and the management of these forest resources has important implications for the quality of urban life. A large number of groups have a significant interest in the management of trees and forests in built environments, including developers, homeowners, retailers, employers, municipalities, arborists, and other professionals. This paper outlines the challenges faced by those who manage trees and forests in built environments, and recommends improvements in the ways that we approach the management of these highly valuable resources. Included are discussions of tree values, interactions of trees and the built environment, construction, tree losses, and the lack of critical information about the interactions among trees, people, and the built environment.

Trees in Built Environments are Valued by People

Trees provide people with a wide range of important benefits, including a more pleasant, healthful, and comfortable environment in which to live, work, and play; and improvements in individual and community well-being (1). Of particular significance are the trees and forests with which people interact on a regular basis, including those near their homes, along streets, in parks, and where they work. The values of individual trees often depend on their size and location, with the most valuable trees — often mature ones — in close proximity to buildings, roads, recreation facilities, or utilities. These trees can positively influence people's feelings, attitudes, moods, and behaviors (2,3,4,5,7,9,10). Trees may occupy very special places in individual, family, and community life; and are often seen as powerful symbols. The following quotes from two individuals who participated in a tree care training course in Chicago illustrate the significance of trees in residential areas (18,19).

- "They [trees] offer shade and a sort of grace to our streets. I enjoy bird

USDA Forest Service, North Central Forest Experiment Station, 845 Chicago Avenue, Suite 225, Evanston IL 60202-2357

watching and can right out of my living room window because of great trees on my street (and in the community)."

- "When my family moved into our new home, my father planted 3 flowering trees (2 sour cherries and 1 plum), one for each of the three children. Late in the winter, my mother cut a few branches and within days, we'd be enjoying beautiful blooms inside."

Many individuals have a strong appreciation of and intense feelings toward trees they encounter in their everyday lives. These strong personal ties to trees may approach a spiritual nature. The following quotes from two visitors to Morton Arboretum near Chicago illustrate some of these feelings (4).

- "There is at once strength, and form in these native trees that create a potent force and magic in the area."
- "A feeling of quiet, peace, and order arises within me."

In their evaluations of street trees, householders tell us that the trees make significant contributions to the residence, neighborhood, and community; including "enhances look of yard and home," "increases property value," "increases sense of community," and "increases sense of home and family" (11). Residents of inner-city Detroit reported that trees along residential streets were the most important to them, followed by trees in city parks, front yards, back yards, downtown areas, industrial areas, and parking lots (6). When asked to rate the relative importance of government provision of "tree areas," they ranked "tree lined streets" first, followed by "open park areas," and "wooded areas" (6).

Given their large size and distinctive character, trees often shape the landscape. They frame and enhance buildings and associated greenspace, and provide a distinctive character and identity to the developments, many of which are named after trees. Buyers are increasingly attracted to new developments where pains are taken to minimize environmental damage so as to provide quality home, leisure, or work environments. This often requires innovative development strategies like set asides, conservancy areas, and cluster development to preserve forest environments, as well as saving individual trees.

Evidence of the high values of trees in built environments can be summarized by the following (1,2,3,4,10):

- Higher sale prices for residential lots that have trees on or near them.
- Higher rental prices for apartments and offices with attractive views of trees and forests.
- People's stated preferences for having trees in a wide range of single and multiple family residential environments, parks, trails, and street corridors.
- The positive responses of homeowners, park users, and office workers to the forest environments they experience.
- The positive responses of shoppers and business operators to beautifi-

cation efforts (including tree planting) in shopping areas.
- The heroic (although often misguided) efforts by developers and landowners to protect, restore, and enhance trees on building sites (i.e., barricades, walls, cribs, soil treatments and amendments, etc.).
- The substantial investments in tree planting on newly-developed sites.

Interactions Among Trees, Buildings, and Utilities are Critical

Some of the most highly-valued trees are found in close proximity to buildings, facilities, and utilities. The combination of trees with buildings and other developments often creates highly preferred environments. Some homes are built around existing trees on a lot or carefully positioned with respect to them. Trees along streets and utility corridors can mask utility wires, equipment, unsightly views, etc. Trees are a key component of a wide range of parks and other recreation areas, often providing significant enhancement of facilities and other resources (5). To effectively manage these complex environments, it is critical that the interactions between trees and other resources are given careful consideration in site planning, design, management, and maintenance.

When considering trees in built environments, it is critical to take a long-term view of the interactions between them. It is important to accommodate tree growth in size over time, both above and below ground. There may be important needs to change and maintain buildings, facilities, and utilities over time. Without appropriate planning, lack of space often becomes a critical problem with trees in built environments.

Trees are strongly affected by development. Construction influences critical dimensions of the microclimate in which the tree grows, as well as the growing space available to the tree above and below ground.

There are social implications of tree management in the development process as well. In some instances the "threatened" or "vulnerable" status of trees and forests in a built environment enhances the emotional attachment that many urbanites have for them and increases their desire that the trees survive (2).

Trees may also directly affect what is built near them. Trees may grow into utility lines and against buildings; and they may block views and shade solar collectors. Tree roots may grow into sewers and other drainage structures; and also damage sidewalks, trails, and roads. Falling trees or parts of trees may damage buildings and facilities.

Important interactions between trees and other resources are reflected in the problems that householders report for street trees (13,14,15). These problems, in order of decreasing significance for trees in the Chicago area, include (12): falling leaves in autumn; fruit, nuts, sticks, or pods fall from the tree; branches or suckers grow from base or roots; insect damage to tree; attracts annoying insects; diseases on tree; roots too close to the surface; falling limbs; leaves fall continuously throughout summer; flower parts fall from tree; sap drips from tree; blocks sun so lawn won't grow; causes allergies; roots clog sewers; sidewalk damaged by tree roots; blocks view; makes street or yard dark; reduces personal safety by limiting visibility. Most of these problems emerge only because the trees are in a built environment and would disappear or at least be far less significant for trees elsewhere. The problems might also be more serious for trees that are closer to homes and other buildings.

Significant interactions between trees and the built environment are not confined to new developments. Important problems may be encountered with older neighborhoods where trees have long been established and residents are attached to them. For example, the following example of a recent sidewalk replacement in a Milwaukee suburb was related to the author by a colleague:

> "When the sidewalks were replaced, many of the neighbors were agitated by how poorly they felt the contractors treated the giant elms on the parkway strip in front of their homes. Removal of the old sidewalks exposed the very large and extensive root systems of the elms. Witnessing the contractors cut the roots away to lay the new sidewalks was a wrenching experience for residents, many of whom had lived on the block for 30+ years and had gone through the past threats of Dutch elm disease. Several neighbors refused to have their sidewalk repaired because they feared it would damage or kill the trees, and one resident made contractors wary of their actions by calling city supervisors and by standing out on her front lawn videotaping the entire operation."

Construction is a Key Point in Present and Future Interactions

Construction is a key point in the interactions among buildings, facilities, utilities, and trees, particularly when construction is in close proximity to or otherwise affects large trees. The type of construction and how it is carried out can have a significant influence on the future growth and development of existing trees, as well as those planted after construction.

Decisions concerning the construction process on forested sites may involve high value trees and the possibility of significant alteration of construction and subsequent maintenance activities and costs. Critical choices include which trees will be retained on the site and what efforts will be undertaken to provide for tree growth and development. Increased construction and maintenance costs to save trees must be weighed against the likelihood of successfully preserving the trees.

In some instances the best solution to conflicts with trees in a built area is to avoid or relocate construction. This may include putting liners in damaged sewer pipes rather than replacing them; tunneling under trees to put in utility lines; foregoing some sidewalk replacements; or relocating construction activities away from the trees, their roots, and areas that will affect drainage and other aspects of the microclimate for trees.

Unsuccessful efforts to preserve trees on a site may generate problems over an extended period of time, such as damage from deteriorating trees; and the high cost of removing trees in close proximity to buildings, facilities, and utilities. The personal anxiety of residents associated with a prolonged decline of an important tree may also be significant, particularly when the trees have high symbolic values.

Construction alters the site for the future, setting the stage for growth and development of existing and newly-established trees. Construction-induced alterations in the site may range from readily apparent changes in soils and microclimate to subtle but very important changes in drainage. These changes can have a profound impact on the future development of trees and forests on the site and nearby areas.

The future growth of newly-planted trees on the site may have important implications to owners and users. These trees will make up the future urban forest. Tree planting has significant meanings to homeowners and the community, including demonstrating commitment to the future, changing the landscape, and improving the environment. The significance and success of tree planting is often increased by homeowner involvement in the effort (16,17).

Loss of Trees is Often Critical to Owners and Users

Loss of trees in built areas seldom goes unnoticed, whether it is removal of trees during construction or their decline and death after construction. The following quotes from two individuals enrolled in a tree care course in Chicago illustrates the human loss that can be associated with special trees (18,19).

- "There was a huge oak tree at the bottom of our garden on our neighbor's side of the stream. One day when I was eleven years old I came home from school and it had been cut down. Twenty six years later there is still a wound in the sky."
- "...my father had planted these trees from twigs, and they were absolutely gorgeous. And then the city came and cut them down .... I was daddy's little girl. So it was hard for me to accept that. Death has been something in my life very young. I could have understood that if the tree was dying. Then, we can remove it. But it was beautiful. So it was hard. It was hard for me to deal with."

Loss of trees in the development process may generate considerable concern in the community and could result in stronger resistance to subsequent development, or calls for increased controls on developers and the development process. Home or business owners who selected tree covered sites for their buildings may be very upset over the loss of trees during construction. Nearby owners may also be upset over changes in their landscape brought about by the loss of trees on nearby areas.

The author has in recent years received two highly-memorable phone calls for critical and urgent assistance from families wanting to know how to propagate trees that were located on the grounds of their family home. In both instances the home was being sold and family members wanted a special tree to live on in their yard.

Post-construction loss of trees that were damaged or their environments altered by construction can be a serious problem. This can result in a long process of decline that is accompanied by a considerable amount of anxiety, uncertainty over appropriate management strategies, and grief (20). Similar long-term problems can emerge with the growth and development of trees planted on construction sites where insufficient attention was given to the requirements for their growth and development. This might include inappropriate soils and drainage.

Hull's work on the human loss associated with trees lost in Hurricane Hugo documents the profound emotional ties between people and urban trees and forests. Residents saw Charleston's trees as symbolizing spiritual values, personal memories, reminders of the past, preservation, and endurance (8).

Lack of Information is a Critical Problem

Many problems with trees in a built environment have their origins in lack of information to guide important decisions. Most people, homeowners, or builders/contractors do not realize what a tree needs to survive, how easily trees can be damaged in the building process, what can be done to help save a damaged tree, or how long it takes for a new tree to grow to a significant size.

There is also a great deal of uncertainty over the extent of the damage that trees can cause to nearby resources. Several years ago Steve Bylina, Chief Forester for the City of Chicago offered that the two major questions with which he needed help from researchers were 1) Do trees damage sound sewers (i.e., enter sewers without breaks in them)? and 2) When is it necessary to remove a tree because it posed a risk to life and property? Both questions figure prominently in tree removal decisions in built areas.

There is limited understanding of the significant values that individual trees have to urban people. People to tree bonds may be very strong, and loss of individual trees during or after construction have serious consequences to individuals and communities that are significant and lasting (4).

We also know very little on how to work to restore trees that have been damaged or otherwise influenced by construction activities, or when to give up and remove them.

Conclusions

The management of built environments poses important decisions that concern trees, elements of the built environment, and interactions between them. These decisions have long term implications for the well-being of individuals and communities. With appropriate planning and management, trees and elements of the built environment complement each other in providing highly desirable environments for a number of activities. But without appropriate planning and management, serious problems can emerge with 1) trees posing major problems to the functioning of elements of the built environment, and 2) the built environment posing critical problems for the survival and growth of trees. Tree and built environment interactions are critical considerations in construction planning; but should also be a integral part of the planning for site management, maintenance, and subsequent tree planting.

Given the high values that people often place on large trees located in close proximity to buildings, facilities, and utilities; significant attention should be given to decisions concerning the retention and maintenance of those trees on built sites. These trees often have very high values and it will take a very long time for newly-planted trees to grow to a size that will replace them. When construction is considered on forested sites, a wide range of options should be considered in the planning process, which should include relocation of construction away from critical trees and forests, as well as alternatives to construction. It should also be recognized that not all trees can be saved during the construction process, and that attempts to save trees that have been damaged or their environments radically altered can be unsuccessful and lead to serious problems down the road. It is critical to look beyond the construction process to the years ahead to provide for the growth and development of existing and newly-planted trees as well as the maintenance of trees and the built environment. Decisions about trees and building sites must be formed from the best information available about trees, their values, and how trees and people are likely to respond to changes in the construction and maintenance processes.

A number of groups should focus significant attention on trees in built areas.

Developers should be concerned with the trees in building sites since they can increase the value of developments and the speed with which they are sold or occupied. Sound management of trees and associated resources during the development process can enhance public perceptions of development and developers, with long-term implications for the development process.

Homeowners should be concerned with trees since they increase the value of their residence, and can provide a wide range of benefits. Trees can also increase the resale value of residential property.

Retailers should be concerned with trees since they can contribute to an attractive environment that will draw shoppers and encourage them to stay longer.

Employers should be concerned with trees since a view of trees from a window can enhance worker satisfaction and productivity.

Municipalities should be concerned with trees throughout their community since they can increase property values and retail sales, and hence the tax base and perhaps tax revenues. The trees and forests in a community often add a great deal to the "character" of that community.

Arborists and urban foresters should be concerned with providing information that will help guide the critical decisions that are faced concerning trees in built environments. These include the requirements of trees and their likely growth and development under alternative construction and maintenance regimes, as well as the benefits of trees and forests in built areas. It is important that we are able to identify the change in benefits that is likely to occur when existing large trees are removed and new smaller trees are planted.

Literature Cited

1. Dwyer, J.F., E.G. McPherson, H.W. Schroeder, and R.A. Rowntree. 1992. Assessing the benefits and costs of the urban forest. J. Arboric. 18:227-234.
2. Dwyer, J.F. and H.W. Schroeder 1995. The social aspects of urban forestry: Toward a deeper understanding of values. pp. 66-78. **In** Partnerships for livable places: Proceedings of the National Urban and Community Conservation Symposium, National Association of Conservation Districts.
3. Dwyer, J.F. and H.W. Schroeder 1994. The human dimensions of urban forestry. J. Forestry 92:12-15.
4. Dwyer, J.F., H.W. Schroeder, and P.H. Gobster. 1991. The significance of urban trees and forests: Toward a deeper understanding of values. J. Arboric. 17:276-284.
5. Dwyer, J.F., H.W. Schroeder, J.J. Louviere, and D.H. Anderson. 1989. Urbanites willingness to pay for trees and forests in recreation areas. J. Arboric. 15:247-252.
6. Getz, D.A., A. Karow, and J.J. Kielbaso. 1982. Inner city preferences for trees and urban forestry programs. J. Arboric. 8:258-263.
7. Hull, R.B. IV. 1992. Brief encounters with urban forests produce moods that matter. J. Arboric. 18:322-324.
8. Hull, R.B. IV. 1992. How the public values urban forests. J. Arboric. 18:89-101.
9. Kaplan, R. 1993. Urban forestry and the workplace. pp. 41-45. **In** Gobster, P.H. (Ed). Managing urban and high use recreation settings. Selected papers from the Urban Forestry and Ethnic Minorities and the Environment Paper Sessions at the 4th North American Symposium on Society and Resource Management. General Technical Report NC-163. USDA Forest Service, North Central Forest Experiment Station, St. Paul MN.
10. Schroeder, H.W. 1989. Environment, behavior, and design research on urban forests. pp. 87-117. **In** Zube, E.H. and G.T. Moore (Eds). Advances in Environment, Behavior, and Design, Vol 2. Plenum Publishing Corporation, 1989.
11. Schroeder, H.W. and S.R. Ruffolo. 1995. Householder evaluations of street trees in a Chicago suburb. Unpublished manuscript. USDA Forest Service, North Central Forest Experiment Station, Evanston IL.
12. Schroeder, H.W. and S.R. Ruffolo. 1993. Householders' evaluations of street trees in suburban Chicago. pp. 68-74. **In** Gobster, P.H. (Ed). Managing urban and high use recreation settings. Selected papers from the Urban Forestry and Ethnic

Minorities and the Environment Paper Sessions at the 4th North American Symposium on Society and Resource Management. General Technical Report NC-163. USDA Forest Service, North Central Forest Experiment Station, St. Paul MN.
13. Sommer, R., P.A. Barker, H. Guenther, and K. Kurani. 1989. Householder evaluation of two street tree species. J. Arboric. 15:99-103.
14. Sommer, R., H. Guenther, and C.L. Cecchettini. 1992. A user-based method for rating street trees. Landscape Res. 17:100-107.
15. Sommer, R., H. Guenther, and P.A. Barker. 1990. Surveying householder response to street trees. Landscape J. 9:79-85.
16. Sommer, R., Learey, F., Summit, J., and M., Tirrell. 1994. Social benefits of resident involvement in tree planting: Comparisons with developer-planted trees. J. Arboric. 20:320-328.
17. Sommer, R., Learey, F., Summit, J., and M. Tirrell. 1994. The social benefits of resident involvement in tree planting. J. Arboric. 20:170-175.
18. Westphal, L.M. 1992. Why trees? An exploration of urban forestry volunteers' values and motivations. M.A. Thesis, Department of Geography and Environmental Studies, Northeastern Illinois University, Chicago IL.
19. Westphal, L.M. in press Birds do it, Bees do it, But why do volunteers do it? A look at motivations. **In** Proceedings of the Midwest Oak Savanna Conference. Feb 20, 1993. U.S. Environmental Protection Agency, Chicago.
20. Windle, P. 1992. The ecology of grief. BioScience. 42:363-366.

Street Reconstruction and Related Tree Decline

Robert W. Miller and Richard J. Hauer

Street trees, streets, curbs and sidewalks are all important elements of urban infrastructure, but trees and hardscapes can come into conflict with one another. As trees mature, roots and the root collar can displace curbs and sidewalks, especially if the trees are fast growing and have a large mature size, and growing space is limited. In interest of public safety, roots are severed, and curbs and walks replaced, but this can cause considerable damage to tree roots (1).

Communities can readily calculate the damage by trees on infrastructure by summarizing construction costs. For example, 22 communities in California incurred an average annual cost of $27,000 for sidewalk repair (2). However, the effect of street, curb and/or sidewalk reconstruction damage on tree survival, health and value is not well known. Street reconstruction is an ongoing process that affects street trees, even when trees are not the cause of the need for repair activities. The objectives of this study were: a) to determine if a relationship exists between street reconstruction projects and tree survival, and condition; b) to determine if there were differences in the condition of damaged trees in relation to width of the tree lawn, tree species, and tree diameter; and c) to assess the economic impact of construction activities on tree value (3).

Methods

The city of Milwaukee, Wisconsin was selected as the study site because a street tree inventory of approximately one third of the city in 1979 assessed tree condition using the Council of Tree and Landscape Appraisers (CTLA) method. Records of construction activity in the inventory area were obtained for the years 1981 through 1985. These activities consisted of street widening and curb setback, curb and sidewalk replacement, curb replacement, and sidewalk replacement. Data were collected in 1989, allowing four to eight years for tree damage to become evident.

One hundred projects (20 per year) were randomly selected, and one block sections were randomly selected from each project. Construction damage was prioritized by the level of disturbance in the following descending order; street widening and curb setback, curb and sidewalk replacement, curb replacement, and sidewalk replacement. Using these criteria, fifty blocks (10 per year) were selected based on those with the highest level of

Dr. Miller is a Professor in the College of Natural Resources, University of Wisconsin at Stevens Point, WI 54481 and Mr. Hauer is a Graduate Student, Forestry Department, University of Illinois, Urbana, IL 61801

disturbance to existing trees. These blocks were examined, and 25 were chosen that had the greatest species diversity. Control (undamaged) trees were located on the nearest block unaffected by construction activity. For both construction and control blocks the first 25 trees per block were selected.

The following data were collected for each tree sampled (controls and treatments); diameter at 4.5 feet (1.4 m) above ground, species, tree lawn width (curb to sidewalk), and CTLA tree condition rating (100, 80, 60, 40, 20, and 0). The type of construction activity was recorded for trees affected. The inventory from 1979 was examined to determine if each existing tree was a survivor, replacement, or an addition to the population after 1979.

A z-test was used to compare tree survival from 1979 to 1989 for construction versus control blocks. Chi square was used to test for differences in tree condition between construction and control blocks in 1979 and 1989. Analysis of Variance (ANOVA) one way and two way procedures were used to test condition of trees surviving in 1989 versus tree lawn width, tree species, and tree diameter.

Results

A total of 989 trees were sampled, with 510 occurring on construction blocks and 479 on control blocks. Of the 989 trees sampled, 670 survived from 1979 to 1989, 175 trees had been replaced sometime during the period, and 144 trees were new to the population since 1979. Fifteen species of trees were included in the sample, but only green ash (*Fraxinus pennsylvanica*), honeylocust (*Gleditsia triacanthos*) and Norway maple (*Acer platanoides*) were present in sufficient numbers for species analyses.

Tree condition and survival. Chi square analysis found no significant difference in tree condition between construction blocks (77.2%) and control blocks (77.7%) in 1979. However, a significant difference (0.05) was found between construction blocks (71.2%) and control blocks (76.7%) for surviving trees in 1989 (Table 1).

Between the years 1979 and 1989, 81.4% of the trees on control blocks survived and 77.3% of the trees on construction blocks survived. These survival rates were significantly different (0.10) using the z-test (Table 2).

Tree lawn width. Significant differences were found in tree condition relative to tree lawn width when construction damaged and control trees were compared (Table 3). ANOVA one way tests were used to compare tree condition and lawn widths within construction and control blocks and ANOVA two way tests were used to compare tree condition and lawn width independent of construction and control blocks. As lawn width decreases tree condition decreases for both construction and control trees, but it decreases more on blocks where construction occurred.

Tree species. Significant differences in tree condition between construction and control blocks were found using the ANOVA two way proce-

Table 1. Percent tree condition of street trees damaged by construction between 1981 and 1985 versus undamaged trees.

Evaluation date	Percent tree condition Construction	No construction
1979	77.2a	77.7a
1989	71.1a	76.7b

Means in same row followed with a different letter are significantly different (.05% level) using a Chi-square test.

Table 2. Percent tree survival from 1979 to 1989 for street trees damaged between 1981 and 1985 and undamaged by construction.

	Number of trees		Percent*
Tree status	1979	1989	survival
Damaged	432	334	77.3
Undamaged	413	336	81.4

* Percent survival significantly different (.10% level) using a z-test.

Table 3. Reduction in tree condition percent of construction damaged and undamaged street trees by tree lawn width.

Lawn width	Tree condition (%)	
(feet)	Undamaged	Damaged
<4	75.8ab	66.0a
4-7.9	74.8a	69.6a
8-9.9	80.4c	73.0ab
>=10	78.2bc	74.2b
Average	76.7	71.1

Means in same column followed with a different letter are significantly different (.05) level) using Student-Newman-Keuhls'.

dure for green ash, honeylocust and Norway maple. However, no significant differences were found in tree condition for these species when construction and control blocks were compared using the ANOVA one way procedure.

Tree diameter. Tree diameters were grouped into three diameter classes (<7 in., 7-12 in., and >12 in.). ANOVA one way procedure revealed no significant difference in tree condition on construction blocks.

Discussion

Reconstruction of streets, curbs and sidewalks were found to be significantly related to both tree survival and tree condition. Construction damaged trees had 22.7% mortality four to eight years following construction projects, while trees not affected by construction had 18.6% mortality for the same period. Tree condition did not change significantly for trees not affected by construction, but declined 6.1% for trees affected by construction.

Width of the tree lawn was found to significantly influence tree condition of both treatment and control trees, with trees on narrow lawns having lower conditions. Additionally, trees were found to have even lower condition ratings on narrower lawns if they were affected by construction projects.

Although construction damage was found to negatively influence tree survival and condition, this decrease in condition was independent of tree species and tree diameter. The lack of difference is tree species is not surprising, as green ash, honeylocust and Norway maple are species that are widely regarded as highly tolerant of harsh urban conditions and site disturbance. Lack of significance between tree diameter and tree condition on construction blocks was surprising, especially when tree lawn width was found to have a negative influence on tree condition. However, the trees in the study area are relatively young, as they are mostly American elm (*Ulmus americana*) replacements that were planted in the late 1960s and 1970s.

Economic implications. Assuming an average value of $1,100 per tree in the City of Milwaukee, the total population of 200,000 street trees has a value of $220 million. Approximately 3% of the street tree population was subjected to construction projects each year of the study period (1981-1985). This represents 6,000 trees damaged per year with a value of $6.6 million. Subtracting for a CTLA condition reduction of 6.1% this is an annual reduction in tree value of $521,500. Tree mortality associated with construction damage was 4.1% higher. This higher mortality rate of 6,000 trees damaged per year is an additional loss of 246 trees with a value of $270,600. The total annual loss in tree value in Milwaukee due to construction related decline is $792,100 ($521,500 + $270,600).

Acknowledgments

The authors wish to thank Kenneth Ottman and Robert Skiera (retired), Milwaukee Bureau of Forestry for involvement and technical assistance in this project.

Literature Cited

1. Hamilton, R. W. 1988. Significance of root severance on performance of established trees. J. Arboric. 14: 285-292.
2. Hamilton, R. W. 1984. Sidewalk/curb-breaking tree roots. 1. Why tree roots cause pavement problems. Arboricultural J. 8: 37-44.
3. Hauer, R. J., R. W. Miller, and D. M. Ouimet. 1994. Street tree decline and construction damage. J. Arboric. 20(2): 94-97.

Soil Aeration Systems: Do They Work?

Karen Smith, Donald Ham, Ansel Miller, and Timothy Chesnut

Documented successes of soil aeration system use in the field appear to conflict with results of a controlled, scientific investigation on the effect of fill on tree welfare. In this two-year study of white pines subjected to 20 cm of fill and a pavement-like surface cover over the root zone, conditions under fill alone were not severe enough for any "improved" effect to be noticed from the use of an aeration system with fill. Soil/atmosphere air exchange was only slightly altered and tree physiological processes remained largely unchanged. These results underscore the need for further quantitative studies of conditions created by various fill and paving procedures to better ascertain the usefulness of elaborate and expensive aeration systems.

Soil aeration systems are routinely recommended by urban tree care experts when construction plans call for raising the grade around a tree. The theory behind these systems of ventilation pipes and porous fill is that they facilitate better gas exchange between the root zone and atmosphere than fill soil alone. Gas exchange is essential to tree function because roots, during the process of respiration, utilize O_2 present in soil macropores and in turn, release CO_2. Under normal conditions, most soil/atmosphere exchange of CO_2 for O_2 occurs when water saturates the soil macropores and flushes soil air out. As water drains from the saturated soil, new air is drawn into the macropores. Anaerobic conditions develop when exchange is inadequate. This causes roots to lose their ability to absorb nutrients and water and to lack the vigor necessary for new soil exploration. Above ground, photosynthesis rates decline, stomata close, and shoot growth slows. Studies have shown that this deterioration of tree health occurs when root zone O_2 content falls below 10% (2).

Compacted soils, or those covered by materials such as asphalt and soil fill, present a barrier to water infiltration and consequently to gas exchange within the root zone. Yelenosky (4) compared soil air composition under fill and paving to that of an undisturbed site. In the undisturbed forest, soil air consisted of no less than 18% O_2 and no greater than 2% CO_2. In contrast, the soil surrounding trees under clay fill were very poorly aerated, with O_2 contents as low as 1% and CO_2 contents over 20%. Anaerobic conditions also developed rapidly in soils subjected to compaction and asphalt paving, with O_2 content dropping from 20% to 4% in two weeks.

Soil aeration systems, consisting of various configurations of piping and porous fill, have been proposed as a means of maintaining air circulation at the original soil surface under nonporous surfaces or deep fills (1,3). Harris

Department of Forestry, 261 Lehotsky Hall, Clemson University, Clemson, SC 29631

(1) recommended an aeration system with the following components: a tree well wall encircling the tree trunk to prevent fill from coming in contact with the base of the tree, load-bearing geotextile fabric spread on the original soil surface to lend even support to the weight of the fill, a horizontal grid system of perforated pipes connected together and vented to the tree well and final fill surface, a layer of gravel fill surrounding the pipes, and finally a second layer of geotextile fabric between the gravel and soil fill to prevent the layers from mixing.

Scientific and Case Study Results

Individual cases of tree survival following grade increases in which soil aeration systems were incorporated, are often held up as testament to the effectiveness of these systems. The number of these successes paints a convincing picture. However, these soil aeration systems have not been subjected to scientific scrutiny. Other urban tree care measures, such as cavity filling and wound painting, were also founded on "conventional wisdom" and widely accepted by a public that felt that doing "something" was an improvement over no treatment at all. Just as scientific study proved those techniques ineffective, research may show aeration systems to be equally questionable.

A two-year, controlled field study of aeration systems was conducted on 11-year old white pines (*Pinus strobus*) growing in a plantation near Clemson, South Carolina. Trees without fill were compared to those whose root zone grade had been raised 15 to 20 cm by 6.8 metric tons of fill. Clayey subsoil fill, having no components of an aeration system, was compared to crushed rock fill (Figure 1), with and without aeration piping. All fill treatments contained the load-bearing geotextile base recommended by Harris (1) and a top surface cover of water-repellent geotextile and tar paper, intended to simulate the water-shedding properties of asphalt paving.

Anaerobic conditions under the fill treatments were less severe than anticipated (Table I). Soil fill reduced O_2 and increased CO_2 content slightly, but not nearly enough to produce concentrations unfavorable to tree growth and development. Soil aeration under rock fill, with and without piping, was equivalent to that of trees with no fill or covering.

Physiological processes of the trees were largely unaffected by the fill and cover treatments. Treated and untreated trees were equally active photosynthetically and produced similar amounts of shoot growth. Measurements of predawn water potential of the foliage showed that treated trees received an adequate supply of water in spite of the water-shedding tar paper which covered their root zones. Predawn potential, an indicator of root zone water availability, ranged from -0.3 to -0.9 MPa throughout the growing season, but was consistent among treated and untreated trees on individual days of measurement. Water was not a limiting factor, as some of the highest photosynthetic rates appeared on days when predawn potential was lowest.

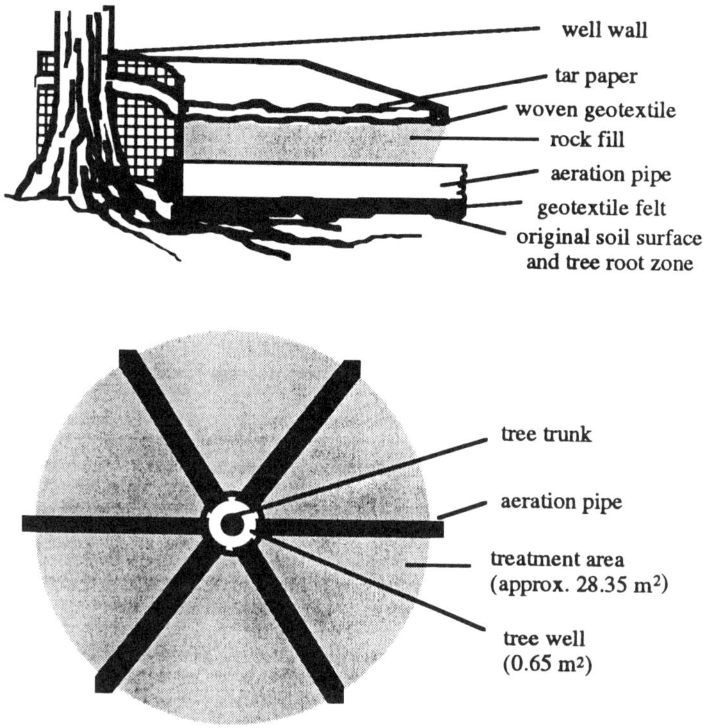

Figure 1. Cross section and plan view of the fill treatment containing crushed rock and aeration pipes.

Table 1. Average percent soil air composition by volume within the root zones of *Pinus strobus* treated with crushed rock or soil fill.

Soil air constituent	Percent by volume		
	Control	Rock fill	Soil fill
Carbon dioxide	0.25	0.26	0.58*
Oxygen	20.60	20.47	20.18

*denotes significant difference from the control.

Discussion

Why did the fill treatments of the controlled study not produce the severe conditions observed by Yelenosky (4)? As a simulation of construction site conditions, this study differed from actual situations in several important ways. Study trees were treated individually with islands of fill and "paving" surrounding them. Hence, the study trees may have gained some access to air and water at the outer edges of the treatment areas, while a tree located within a vast parking lot would have little advantage from edge effects.

Another confounding factor observed on construction sites, but not addressed in this study, is the effect of soil compaction on root zone aeration. Fill materials were placed on load-bearing geotextile fabric with a front-end loader which stayed outside of the treatment areas. Consequently, significant compaction did not occur. Soil bulk densities beneath the fill were low—averaging 1.08 g/cm^3 with pore space comprising 59% of soil volume. Although similar precautionary measures can be taken in actual practice, soils are often so compacted prior to fill application by other construction activities that soil aeration is already limited. This was probably the case in Yelenosky's study where measurements were taken in soils under fill layers that had been rolled and packed down with grading equipment in preparation for asphalt paving.

Confounding factors and the limited scope of this study make it difficult to give conclusive statements regarding the usefulness of soil aeration systems. Further studies must first be designed and conducted which address the following factors: varying species tolerance to poor soil aeration, effects from greater depths and different types of fill, and soil compaction's role in aeration problems. The results thus far seem to indicate, however, that fill may not be as large of a culprit in creating poor soil conditions and tree decline as the other damaging activities that normally accompany it. Perhaps more important than the actual soil aeration system in insuring a tree's survival following grade changes is the increased level of care and protection from damage afforded to a valued tree at the construction site.

Acknowledgment

This project was funded in part by grants from F.A Bartlett Tree Experts and the International Society of Arboriculture Research Trust. The author gratefully acknowledges the guidance of Drs. Ansel Miller and Donald Ham, with whom the initial concept for the study originated. Thanks also go to Drs. Judy Caldwell and Larry Nelson for providing additional advice and resources, and to Kathy Brock for her assistance with the field work.

Literature Cited

1. Harris, R.W. 1992. Arboriculture: Integrated Management of Landscape Trees, Shrubs, and Vines, 2nd Edition. Englewood Cliffs, New Jersey: Prentice Hall.
2. Kozlowski, T.T. 1985. Soil aeration, flooding, and tree growth. Journal of Arboriculture. 11(3): 85-96.
3. Schoeneweiss, D.F. 1982. Prevention and treatment of construction damage to shade trees. Journal of Arboriculture. 8(7): 169-175.
4. Yelenosky, G. 1963. Soil aeration and tree growth. International Shade Tree Conference Proceedings. 40: 127-147.

An Investigation to Assess the Impact of Street Infrastructure Improvements on the Roots of Adjacent Cork Oak Trees

Patricia A. Lindsey, Rob Gross and Bob Milano

In an investigation to assess the potential impact of a street tree improvement project on the root systems of 26 mature cork oaks *(Quercus suber)* on the University of California Davis campus, selected areas of the root systems of two trees were hydraulically excavated. Root size distribution patterns under the asphalt roadway relative to the number of roots in adjacent areas (under decomposed granite and turf) were compared by excavating five 3 m (10') long x 0.30 m (1') wide x 0.60 m (2') deep trenches and the root crown area of each tree. Less than 15% of the total roots uncovered and measured for the two trees were found in the trenches underneath the asphalt. Root size distribution was found to be specific to tree health, surface treatment, and distance from trunk. For both trees, the majority of fine roots were found underneath the irrigated turfgrass and in the root crown area. All large roots extending from the root crown on the curbside angled underneath the road to a depth beyond the point of construction impact. This investigation determined that the street improvement project could proceed with the expectation of minimal impact on the roots of these street trees.

Cork oaks *(Quercus suber)* are a commonly planted tree species on the University of California Davis campus. Native to the Mediterranean region, they are well adapted to the hot dry summers and mild wet winters in Davis. Used primarily as street trees, their large and rounded evergreen canopies provide a dramatic tunnel effect over many campus roadways. The streets in the original part of campus enclose a large open greenspace called the Quad, and were planted with cork oaks over 70 years ago. Now mature, these trees ring the outer edge of the Quad, effectively defining and articulating this historic greenspace, which now serves as the heart of campus life. However, over time the aging asphalt roadways have degraded and been frequently patched. In many cases the curb and gutter is in direct conflict with the substantial trunks of these mature oaks (many with diameters in excess of 1.2 m (4')). The existing antiquated storm water removal system is incapable of providing needed drainage for the greenspace and roadways.

Patricia Lindsey, Landscape Specialist, Environmental Horticulture Department, University of California, Davis, CA 95616
Rob Gross, Consulting Arborist, DendroTech, Calistoga, CA 94515
Bob Milano, Formerly Acting Director of Grounds, University of California, Davis, CA. Now Sports Turf Manager, University of California, Berkely, CA 94720-4426

In the fall of 1993, a renovation plan for these older roads was prepared by the Architects and Engineering Services at the University of California, Davis. Although the renovation plan was specifically developed for the West Quad Roadway, it was intended that this plan serve as a prototype for the subsequent renovation of the remaining roadways around the Quad and possibly for all campus roadways. There were several objectives of this renovation project: 1) to improve the growing conditions for the trees, 2) increase the drainage capability of the Quad and the roadway, and 3) provide a circulation system that more clearly separated pedestrians, bicyclists, and vehicles.

To accomplish this, a new storm water removal system was to be constructed underneath the road with lateral lines installed to facilitate drainage of the adjacent open greenspace. Because of damage to the curbing from the mature trunks, the total road width was also to be reduced an additional 1.2 m (4'). Finally, a new unit block paving detail would be installed on the Quad side of the roadway in order to architecturally define the edge of the Quad and better direct pedestrian traffic.

The protection and continued preservation of these oaks during the renovation of the West Quad Roadway was considered vital to maintaining both the unique character of the Quad and the overall grand stature of the campus urban forest. The UC Davis Grounds Department was given responsibility for the design and implementation of a plan that would minimize harm to the existing trees. To choose the least disruptive plan and address concern over the potential impact that this renovation project might have on the trees lining this street, the following questions needed to be addressed: 1) where were the tree roots - how far out did they extend and to what depth? 2) what percent of the total root system would potentially be impacted by the project? and 3) what would be the probable effect on tree health relative to the percent of tree roots impacted? To answer these questions specific and accurate information on the location of the existing tree roots was needed.

Research on tree root distribution patterns for forest sites and for field production systems has been documented and critically reviewed for applicability to landscape trees (2,8). This research is currently reflected in many tree preservation guidelines - guidelines such as protecting the area underneath the drip line of the tree canopy or staying a distance of 0.3 m per 2.5 cm (1 foot per inch) of trunk diameter from the trunk of the tree. However, it is not at all clear that these guidelines are especially useful for urban trees. The variety and site specificity of urban factors affecting root growth, e.g., soil compaction, the presence of anthropeic materials, stratification, soil infertility, limited soil volumes due to underground infrastructure, impervious paved surfaces and surfaces with a high heat exchange capacity, and site management practices such as irrigation, fertilization and pruning, make the prediction of root distribution highly problematic. Research to date has primarily characterized growing conditions, specifically soil gas concentrations, soil moisture conditions, and soil bulk densities in streetside plantings and other paved areas (5,8,11). Research characterizing root distribution underneath roadways is lacking. For this project, an investigation was undertaken to physically locate and assess street tree root distribution underneath the existing roadway, the planting strip, and the turfgrass to determine if the percent of roots impacted would be acceptable.

Materials and Methods

Tree selection. West Quad Road, with a width of 8.5 m (28'), is one of four roads encompassing the Quad and has a planting of 26 cork oak street trees and 1 valley oak (*Quercus lobata*) (Figure 1). On the west side of the roadway (to the left in Figure 1), these trees had previously been impacted in 1988 by the installation of a trench for a steam line placed within 3.6 m (12') of the trunks. A number of trees on this side are in decline. On both sides of the street the trees are in planting strips, un-irrigated and surfaced with up to 7-10 cm (3-4") of decomposed granite, 4.7 m (15'6") wide on the west side and 8.7 m (28'6") wide on the east side. Many of the trees on both sides of the street have lifted up the street curb and gutter due to the expansion of buttress roots and trunk flare, and in turn have been damaged by this process. Because of the previous impact to the trees on the west side, it was decided not to place the storm water drain line in the centerline of the roadway but instead on the east side of the road. Consequently, this study

Figure 1. A view of the tree lined West Quad Road. Note that the trees on the left side are showing signs of decline. The trenches have been excavated for Tree 1 and the safety barricades are in place while root data are being collected.

addressed the impact to trees on the east side of the road only.

Two trees were selected for root excavation in March of 1994. Project timing prevented the excavation of all the trees on this side, subsequently the trees selected bracketed extremes in size and health. Percent dieback in the crown, leaf color, annual twig growth, diameter at breast height, and radial trunk growth were the factors considered in assessing health. Tree 1 is the larger oak, in good health with a diameter of 1.26 m (41.6"), while Tree 2, with a diameter of 0.72 m (23.8"), is showing evidence of decline. It was assumed that root growth and distribution relative to site conditions could differ depending on tree health.

Trench layout, site preparation, and excavation. The surface treatment immediately adjacent to these trees consisted of asphalt, decomposed granite, and farther away, turfgrass. Two trenches were marked for excavation under the asphalt, two under the decomposed granite and one under the turfgrass. The trenches were positioned parallel to the street curbing, 2.7 m (9') and 5.4 m (18') from the trunk on the asphalt side and 2.7 m (9'), 5.4 m (18') and 8.2 m (27') from the trunk on the decomposed granite/turfgrass side (Figure 2). In addition, the root crown of each tree would be excavated to a 0.6 m (2') depth. It would not be feasible from either a structural or physiological perspective to excavate the entire root zone of any tree, especially a mature one. The assumption being made here is that the roots present in the excavated areas, while representing only a sample of the total root system, accurately reflect the overall trend in root size distribution for the tree relative to surface treatment.

In general, physically locating roots in the field has been accomplished mechanically, carefully excavating the soil by hand or machine. This is a laborious and time consuming method resulting in less than accurate measurements of fine roots. However, the increasing use and refinement of

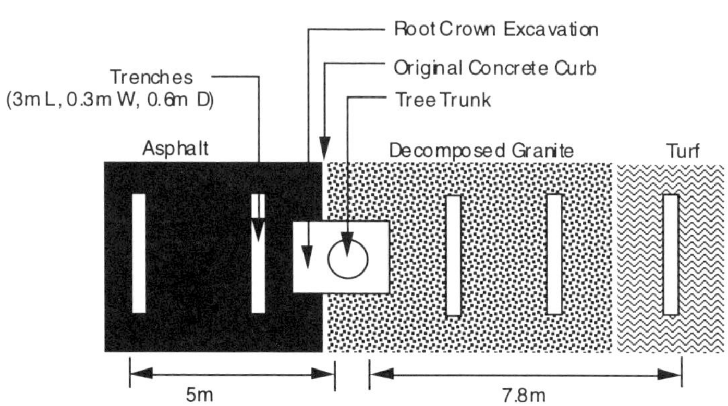

Figure 2. A diagram of the layout and dimensioning of the trenches around each tree relative to adjacent surface treatments (Not to scale).

a relatively non-intrusive technique for examining roots in situ, hydraulic excavation, has shown great promise and was selected for use in this study. Hydraulic excavation involves eroding the soil away from the roots with the application of pressurized water, with the resulting slurry then drained downslope or vacuumed up and deposited for later disposal in a vacuum truck (3). When the slurry is removed the roots are fully exposed. Excavation can be quick and damage to tree roots is generally considered to be minimal. First mentioned in the scholarly literature in 1955, "hydraulic excavation," was used to facilitate the study of root systems of chaparral plants along roadsides (4). Arboricultural applications of this technique include the removal of toxic soil, compacted soil, or soil fill soil from around tree roots, appraising tree hazard potential, boring holes for foundations, diagnosing tree disorders and assessing original soil level (3,6,10).

On March 21, the UC Davis Grounds department removed the two asphalt strips and the curbing in front of each selected tree. Two 1.3 cm (0.5") flexible plastic irrigation pipes with low volume spray heads were used to pre-irrigate the areas for 48 hours prior to excavation. A preliminary study conducted on the campus in the fall of 1994 on the effect of soil moisture content on soil erodibility found that pre-irrigating a silty clay loam to field capacity increased the rate of soil removal by a factor of seven - 0.04 m^3 (1.46 $ft.^3$) of soil eroded per minute for the wetted soil as compared to 0.006 m^3 (0.21 $ft.^3$) of soil eroded per minute for the dry soil.

The five trenches and root collars of each tree were hydraulically excavated on March 23, 1995 (Figure 3). For the soil excavation process UC Davis hydrant water was used, supplied at 4.34 Kg/cm^3 (62 pounds per square

Figure 3. The trenches being hydraulically excavated. A wooden template is laid down and the operator uses this to direct the flow of water and control the dimensions of the trench. As water is applied to each trench, the sump truck is removing the resulting slurry. Note the safety equipment being used by the operator - full waterproof vinyl suit, gloves, hard-hat and safety goggles.

Figure 4. Tree 1 following hydraulic excavation of the five trenches and the root crown. ***(Jack Clark)***

inch). Fire hoses obtained from the UC Davis Fire Department measured 6.25 cm (2.5") in diameter for the supply line and 4.4 cm (1.75") in diameter for the hand line. The nozzle used was a variable flow KK "Thunder Fog" (Task Force Tips, Valparaiso, IN) with a pistol grip. A wooden frame, sized to the outside dimensions of the trenches, was placed on the surface of the ground and water was directed within this frame. This ensured that each trench was excavated to approximately the same surface size and provided a straight edge to direct the stream of water. A Camel brand vacuum truck removed the resulting slurry. Following excavation each trench measured approximately 0.3 m (1') wide by 3 m (10') long, by 0.6 m (2') deep (Figure 4). The entire root collar was also excavated to a 0.6 m (2') depth. The trenches, when not in use, were covered with plywood for reasons of safety and to reduce root drying.

Collection and analysis of root and soil data. Root counts and diameter analysis were limited to five categories: fine (0.5-1 mm), small (2-5 mm), medium (5-10 mm), large (10-20 mm) and very large (>20 mm). Due to the difficulty of separating and accurately counting very fine roots (<0.5 mm), these were not included in the final analysis. Following excavation, the fine, small and medium roots were pruned off one wall of the turf trenches and the curbside of the root crown, separated in the lab and measured with an electronic caliper (Mitutoyo Corp., Japan). The large and very large roots were measured directly in the trenches. For the asphalt and decomposed granite trenches, all measurements were taken directly in the field.

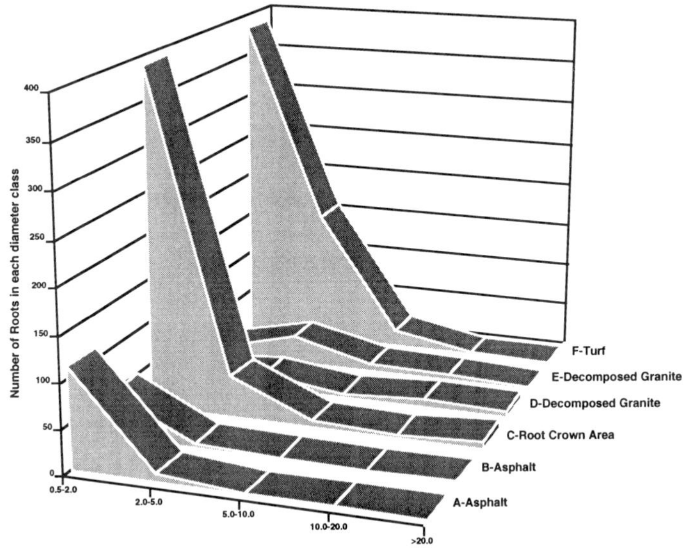

Figure 5. Root size distribution data in the trenches and for the root crown area for Tree 1.

Soil samples in each trench were collected along with a sample of the decomposed granite for textural analysis. Bulk density was obtained using a 5.1 cm (2") diameter, 10.2 cm (4") deep core sampler (AMS, American Falls, Idaho). All soils were dried for 24 hours at 105°C (221°F), then weighed or sieved for particle analysis.

Results and Discussion

For both Trees 1 and 2, the root distribution patterns are graphed relative to surface treatment (Figures 5, 6). While the pattern of root size distribution was the same for both trees, overall, Tree 1 had almost twice as many roots as Tree 2, primarily in the fine, small and very large root size classes. The biggest increase was within the root crown area, increasing from 16.5% of the total to 35.1%. This could be due to the declining health of Tree 2 relative to Tree 1.

For both trees, root size distribution varied considerably both by distance from the trunk and surface treatment. Most of the large and very large root size classes and all of the buttress roots were found in the root crown area. The turfgrass and root crown areas had the highest percentage of fine and small roots. It was also observed that the very fine non-woody roots, those less than 0.5 mm, were concentrated in the turf area and at the root collar of both trees. Given that the turf was frequently irrigated and accounting for rain throughfall at the trunk, this would not be unexpected.

The total percent of roots in the asphalt trenches was 13.9% and 14.6%,

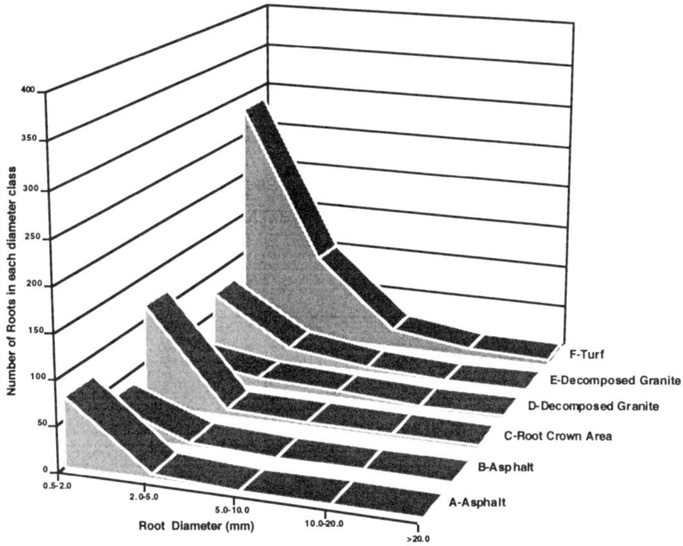

Figure 6. Root size distribution data in the trenches and for the root crown area for Tree 2.

Table 1. Soil bulk densities to specified depths under the different surface treatments. None of the bulk densities obtained exceed the critically limiting bulk density. Each density represents a sample size of one. Given the close proximity of the trees, soil density data was collected for Tree 1 only. Soil texture data were collected for both trees and were identical.

Sample location	Depth	Bulk density (mg/m^3)	Soil texture	Critical bulk (mg/m^3)(1)
Turf	0-4"	1.41	sandy loam	1.70
Turf	4-8"	1.42	sandy loam	1.70
DG	0-4"	1.93	gravelly sand	——
DG	4-8"	1.53	loamy sand	1.75
Road	0-4"	1.60	loamy sand	1.75
Road	4-8"	1.58	loamy sand	1.75
Road	8-12"	1.52	loamy sand	1.75

respectively, for Tree 1 and 2. In the root crown area of each tree it was observed that several large roots angled down sharply underneath the road. These roots did not reappear in either of the asphalt trenches. A visual examination of the utility trench later excavated for the storm water drain, constructed 3.6 m (12') from the original curb and running parallel to that curb to a depth of 1.8 m (6'), confirmed an almost total lack of root development in that area. Because the soil was a loamy sand, aeration would probably not

Figure 7. A close-up of the root crown excavation around Tree 1. Note the line of original soil fill as it appears to be above the point of natural flare. This indicates that soil around these trees may have been raised at some point, perhaps when the decomposed granite was installed. *(Jack Clark)*

have been limiting. Bulk densities decreased slightly with increasing depth and were not critically limiting (1) (Table 1). Given these factors, it was assumed that roots were occurring at much greater depths underneath the roadway and that the impact from the installation of the storm drain would be minimal.

Following excavation of the root crown area, a line of fill above the original root flare was observed (Figure 7). Differences in root morphology between the turf and decomposed granite areas were dramatic (Figure 8). The 8.7 m (28'6") wide band of decomposed granite (DG) is a recent installation. Oaks on the UC Davis campus reportedly thrived for 20-30 years until irrigation was introduced. Irrigation during the dry season can increase the incidence and severity of root rot for this species (6). Approximately ten years ago the turf was removed from under the canopy of these oaks and the DG installed. DG is considered to be an inexpensive solution to the need for a compaction resistant material in the landscape for pedestrian and light vehicular traffic. If used for this purpose, it is important to specify a narrowly graded material. Particle analysis of the DG indicated that it was a very widely graded material, with diameters ranging from a high of 9 mm (3/8") to silt and clay sized particles. This material would compact quite well and result in a slowly permeable surface treatment. In fact, during the wet season, standing water on the surface for weeks at a time had previously been observed. The higher

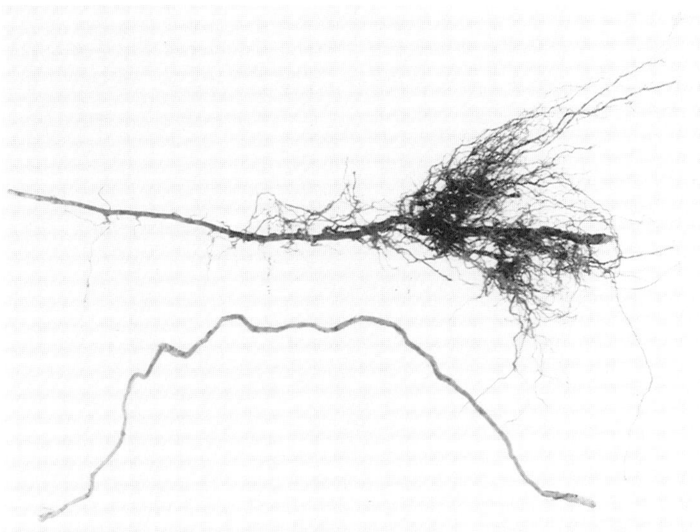

Figure 8. Differences in root morphology under the irrigated turfgrass and under the more compact and poorly draining decomposed granite. ***(Suzanne Paisley)***

soil density and change in soil texture, coupled with the drier soil conditions as compared to the soils underneath the turf, could have affected the change in root size distribution.

Following the excavation and subsequent analysis, it was decided that renovation of the West Quad roadway could proceed as planned. Square tree protection boxes, 1.8 m (6') in height and constructed of 1.9 cm (3/4") thick plywood were placed around the trunks of all trees on both sides of the roadway. The contract called for an arborist to be on the site to monitor construction activity. Construction was completed by September of 1994. The health of these street trees, as of June 1995, does not appear to have been negatively affected by this project.

Knowledge of the location of tree roots is critical to investigating and assessing the impact of infrastructure improvements prior to construction. With this information, changes in the project design can be made that are less destructive to existing trees and more effective protective methods can be specified. Clearly, the hydraulic excavation of carefully located trenches for the two street trees in this project gave an informed understanding of the existing root distribution. The presence of a substantial number of roots less than 0.5 mm in diameter in the turf trench would confirm that hydraulic excavation, given the water pressure used, resulted in minimal root damage. More in-situ research needs to be conducted in order to develop a better understanding of root growth and distribution in urban areas. Hydraulic excavation may facilitate future research in this area.

Acknowledgements

The authors wish to thank Task Force Tips (Valparaiso, IN) for their loan of nozzles and the UC Davis Fire Department for the loan of fire hoses for the hydraulic excavation.

Literature Cited

1. Daddow, R.L. and Warrington, G.E. 1983. Growth limiting soil bulk densities as influenced by soil texture. US Forest Service, WSDG Report, WSDG-TN-OOOO5, 17 p.
2. Gilman, E.F. 1990. Tree root growth and development. I. Form, spread, depth and periodicity. J. of Environ. Hort. 8(4): 215-220.
3. Gross, R. 1993. Hydraulic vacuum excavation: Getting down to the roots. Arbor Age. 13(7): 10-12.
4. Helmers, H., J.S. Horton, G. Juhren, and J. O'Keefe. 1955. Root systems of some chaparral plants in southern California. Ecology 36(4): 667-678.
5. Klein, C. 1980. Characterization of Syracuse, N.Y. streetside soils for urban forestry interpretation. Dissertation, State University of New York, College of Environmental Science and Forestry, Syracuse, N.Y. 111 pp.
6. Miller, R. 1994. Protecting contaminated trees: The Sharon Steel/Midvale Tailings Project. October. pp 33-35.
7. Mircetich, S.M., R.N. Campbell, and M.E. Matherron. 1977. Phytophora trunk canker of coast live oak and cork oak trees in California. Plant Disease Reporter 61(2): 4-6.
8. Perry, T.O. 1982. The ecology of tree roots and the practical significance thereof. J. Arboric. 8(8): 197-211.
9. Van Camp, J.C. 1961. Tolerance of trees to soil environment. National Shade Tree Conference Proceedings 37: 5-19.
10. Watson, G.W. 1990. Tree growth revisited. Golf Course Management. pp. 7-12.
11. Yelenosky, G. 1963. Soil aeration and tree growth. International Shade Tree Conference Proceedings 39: 16-25.

Tree Root Damage From Utility Trenching

Gary W. Watson

Growing, healthy tree roots require water, oxygen, nutrients, low penetration resistance and appropriate rhizoshpere microflora from the soil environment. Deep soils are often dense and lacking one or more of these factors, and surface soils are rich in them. Consequently, the soil environment has a great influence on root growth and distribution.

In a soil profile, the most porous layers are usually near the surface. Water and oxygen occupy these pore spaces. A predominance of micropores, or periodic high soil saturation, can reduce the amount of soil oxygen. This is less likely to occur in the highly porous surface soils. As oxygen is used by roots and other soil organisms, it must be replaced by oxygen that diffuses into the soils from the atmosphere. The highest levels of oxygen will be found near the soil surface, close to the source. Diffusion slows in the deeper soils with less pore space.

Many essential nutrients are released by the breakdown of organic matter near the soil surface. Only the mobile nutrients will move to roots growing deeper. Though roots grow down in response to gravity, other soil environment factors can often have a greater influence. Roots can actually grow up towards the surface soils, if the growing conditions there are more favorable. Soil organisms thrive in the rich surface soils also, and roots depend on many soil microorganisms, e.g., mycorrhizal fungi.

Urban soils often restrict depth of root penetration (Fig. 1). Man-made soil profiles resulting from land development and reshaping often are composed of a compacted clay subsoil layer overlain by a shallow topsoil layer. Roots often fail to penetrate the compacted subsoil, limiting most roots to the topsoil layer, which may be less than 4 inches (10 cm) deep (Fig. 2). Extremely shallow root systems can lead to frequent episodes of stress. For instance, the thin topsoil layer can dry out quickly between rain events. Since infiltration of water into the subsoil is slow, a perched water table can develop in topsoil after rainfall. Alternating stresses from drought and waterlogging hinders fine root development. Such a weakened plant will not be well prepared to endure additional stress from root system disturbance.

The majority of the tree root system is often limited to the upper 2 feet (60 cm) of soil, even on high quality sites. Some species develop taproots as seedlings, but deeper soils soon stagnate taproot growth. This root may persist, but rarely grows large in comparison to trunk size. The prominent root flare of trees is the transition zone between the vertical trunk and the

Dr. Watson is a Root System Research Biologist with the Morton Arboretum, Lisle, Illinois 60532

Figure 1. The shallow spreading roots and lack of a taproot can often be observed when trees topple over.

Figure 2. Roots often spread much further than the branches.

horizontal roots.

Conflicts between trees and underground utilities are common. Service interruptions can easily result from tree planting activities. Installation and repair of underground services often result in extensive damage to tree root systems. Ideally, underground services and trees should be located in separate areas, but in reality, there is rarely enough space in urban areas to accomplish this. Even if primary lines can be installed under the paved street surface where tree roots seldom grow, the numerous service lines to individual homes are almost always within the natural root space of trees.

A single narrow trench less than 2 feet (60 cm) deep can, if located close to the trunk, cut off nearly half of a tree's root system. Stress, decline and death may result, but there is an alternative. The technology exists today to install utility cables and pipes underneath and through the root systems of trees. Specialized equipment can be used for directional boring (also called augering and tunneling) through the soil (Figure 3). The direction of the boring can be controlled around obstacles, such as rocks, roots and other utilities. Little if any damage is inflicted on the root system. With this equipment, there is no need to dig open trenches in the root zone of trees. The increased cost may be balanced by reduced cost of removing and replacing trees killed by trenching.

Even the casual observer can note that utility installation near trees often results in decline or death of the tree. Until recently, there have been no data available to document the effects of trenching on tree survival and vigor. Morell (3) published recommendations for boring distances relative to tree

Figure 3. Directional boring equipment can be used to install underground utilities without disturbing tree root systems.

Table 1. Specifications for boring distance in relation to tree size (adapted from Morel [3])

Tree diameter	Total distance
5-9 in.	10 ft.
10-14 in.	20 ft.
15-19 in.	24 ft.
>19 in.	30 ft.

Table 2. Comparing the costs of boring and removal of trees damaged by trenching (adapted from Morell [3])

	Trenching	Boring
Base project cost	$320,000	$320,000
Boring costs	$0	$32,000
Trees killed	44%	none
Tree removal & replacement costs	$29,000	$0
Actual direct project costs	$349,000	$352,000
Value of lost trees$137,000	$0	

size which were essentially an extrapolation of nursery industry standards for root ball sizes (Table 1). The basic logic is sound, but there may be differences between transplanting and trenching that warrant special consideration.

Transplanting results in root loss on all sides of the tree. Trenching usually causes root loss on only one side. Under some circumstances, trenching may not be as stressful to young trees as transplanting. As trees become larger, they usually are less vigorous, and just as transplanting success is reduced, the ability to withstand root loss from trenching could be decreased as well.

Morell (3) compared the estimated cost of tunneling to the actual costs of removing and replacing trees that died after trenchs were dug to install new water mains near large trees (Table 2). Forty-four percent of the trees were lost. Removal and replacement of these trees cost $29,000. Boring would have added $32,000 to the cost of the project. The net direct cost of boring would have been only $3,000. When the appraised value of the trees that died is taken in to account, boring would have saved $134,000. This points out a major problem with the way construction projects are often organized. Not all of the indirect costs of a project are included in the project budget. If the cost of repair, removal and replacement of damaged trees were part of the project budget, the economy of directional boring would be more appreciated.

Recently, Miller and Neely (2) monitored the survival and vigor of trees along streets after trenches were excavated on one side for new fiberoptic telephone cables (Table 3). When the trench was located closer to the tree than recommended in Morell's specifications (3), trunk diameter growth was reduced. Growth of hackberry (*Celtis occidentalis*) trees was reduced significantly for the duration of the study when the trench was dug within 18 inches (50 cm) of the trunk. The trench was dug on the side of the tree with the largest potential root space, contributing even more to stress. A trench within one meter of the sugar maple (*Acer saccharum*) trees, with only 30 percent of the clearance recommended in the specifications, reduced trunk growth significantly the first year, but surprisingly, not after the first year. There was a severe drought the second year after the trenching and growth of the control trees was also reduced by 75 percent. The controls may have been under as much stress as the trenched trees, thus masking reduced growth from trenching. Growth for both treated and control trees returned to normal the third year. Two years of reduced growth is reasonable for trees of this size, species and amount of root loss. A trench 8.5 feet (2.5 m) from the sweetgum (*Liquidambar striaciflua*) trees (85 percent of the clearance specified) reduced growth slightly, but not significantly. The specified distance for the trench and the actual distance were the same for the honeylocust (*Gleditsia triacanthos*) trees, and there was no growth reduction.

These data show that the closer the trench is to the tree (and the more roots that are presumably lost), the more seriously the tree is impacted. The surprising fact is that none of the trees died as a result of the trenching, but these were not large trees. Averaging 10-14 inches (25 to 35 cm) dbh, they were still young and vigorous. Young trees are known to survive disturbance from construciton and transplanting more readily than old trees. Details of the sites on which trees were growing were not described, but could reveal

Table 3. Effectiveness of boring specifications (3) as demonstrated by data from Miller and Neely (2)

	Celtis occidentalis	Liquidambar striaciflua	Acer saccharum	Gleditsia triacanthos
Trunk diameter (cm)	25	27.5	27.5	35
Specified boring distance (m)*	22.4	3.0	3.0	3.3
Actual trench distance	0.5	2.6	1.0	3.3
Trunk growth reduction				
year 1	--	12%	36%	0%
year 2	49%**	33%	0%	0%
year 3	44%	7%	0%	0%
year 4	48%	0%	0%	0%

* Distance from trunk at which boring should begin (3)

** Reduction in growth was significant ($p<0.05$)

other contributing factors. For example, if the trench was close to the curb, root loss may have been minimal, and if the root space was ample in other directions, root system reduction may be as severe as it might seem from the location of the trench alone.

The Morton Arboretum Study

A study has been underway at The Morton Arboretum for the last four years to investigate 1) the effects of root loss from trenching and 2) the value of crown pruning to compensate for the root loss. To study the effects of severe trenching, the treatments included trenches on one, two, or three sides of the tree, less than one-third of the recommended distance (3) and resulted in approximately 33, 66 and 75 percent root loss(Fig 4). The crowns of one-half of the pin oaks (*Quercus palustris*) were reduced by approximately 30 percent. The trees averaged 12 inches [30 cm] dbh and grew on an undisturbed site with quality soil.

While the experiment is still in progress, several trends have to developed in the data:

- Vigor decreases with increasing root loss
- Diameter growth was less affected than twig growth
- Dieback increases with increasing root loss
- The more severe the root loss, the longer the period of reduced vigor
- Pruning may minimize twig growth reduction associated with root loss.

Stress was evident within one month of the trenching. Tree crowns became chlorotic, especially those trees that lost 75 percent of their root system. Branch dieback was greater in treated than in untreated trees, but not to the extent that the tree crowns were misshapen. The trees looked very much like

Figure 4. Trenches were dug to sever roots in a study of the effects of root damage. In the most severe treatment, trenches were dug on three sides.

stressed, transplanted trees, which can loose as much as 95 percent of their root system, and then recover after a period of stress. This stress period can last until the tree is fully established - up to a full year for every inch dbh in the midwestern climate, .

The trees have been regaining vigor during the four years since the root damage was inflicted (Fig. 5). Twig growth and trunk diameter increase each year, but are still less than the controls for the more severe treatments. Color is improving steadily each year as the root system regenerates.

Other than the trenches, the soil was not disturbed on the site. When the soil was placed back into the trenches it was allowed to settle naturally, and not compacted. Excavations showed that root regeneration was progressing across the trench and into the undisturbed soil. Indeed, the trees are reacting very much like new trees planted on a quality site (Fig. 6).

Morell's (3) recommendations have a great deal of merit. As standard specifications, they will assure a high probability of tree survival if followed faithfully by contractors in the field. However, the more recent information suggests that individual site evaluations by qualified arborists would be even better under many circumstances. Several factors must be considered for each tree to minimize costs of directional boring: percent of roots lost, disturbance history, species, age of tree,vigor of tree, crown size, and expected maintenance.

After assessing the possible root space distribution, and the extent of possible root loss caused by the path of the utility trench, the boring distances

Figure 5. Four years after 75 percent root loss from trenching, the crown is thin and growth is reduced, but improving, and branch dieback is noticable. Transplanted trees of similar size often have the same appearance.

Figure 6. When severed roots can regenerate, the trees may eventually recover from trenching after several years of stress.

might be increased or decreased. For example, if the utility installation must take place on the side of the tree where the majority of the root system is likely to be, trenching up to the standard distance on that side would cause a greater than normal proportion of root loss for that tree, and the boring distance should be increased to compensate. Conversely, trenching near a curb or other obstacle which already prevents roots from spreading might result in so little root damage that boring would not be necessary. Trees with previous root disturbances may require longer boring distances if the previous disturbance was to another part of the root system, or none at all if the trench could go through the area where the roots were already lost from another recent disturbance.

Other concurrent site disturbances are also important. A well-repaired trench as the only stress will probably be less destructive than at a site where there has also been other disturbances, such as soil compaction, a new pavement surface, or drainage changes. A trench which is promptly refilled would be considered a temporary root loss situation, which eventually can be overcome. When the trench is a first step in a grade change, pavement installation, or root barrier installation, the roots will not be able to regenerate and the root system permanently may be inadequate to support the crown. Sites where root systems are restricted unusually close to the surface by heavy soils may warrant longer boring distances because the root system probably spreads further. It may be possible to shorten boring distances slightly on deep soil sites.

Species, age, crown size and vigor will have a major impact on boring distances. So-called "tough" species will often tolerate more root damage than "sensitive" species on the same site. Older trees and stressed trees may be severely impacted by slight or moderate root loss. Broadly spreading trees may require more extensive boring than trees with small crowns. Maintenance practices before the root loss event can affect vigor and the tree's ability to withstand stress, but maintenance afterwards is even more important. If special maintenance, especially watering, can be continued for several years, then the tree may be able to survive the stress associated with greater root loss. The larger the tree, the longer it will need watering during dry periods and the savings on boring may be offset by increased maintenance costs.

Root loss of established trees needs more careful study so that arborists can more accurately prescribe cost effective-protection measures when any kind of construction disturbance is going to occur near trees. Severing of roots and the resulting reduction in the size of the root system supporting the crown is important, but cannot be the only factor involved, since transplanted trees survive consistently with even greater root loss. Work is continuing at The Morton Arboretum to investigate how trees cope with disturbances in the root zone, focusing both on root injury and soil environmental changes.

Literature Cited

1. American National Standards Institute. 1986. American Standard for Nursery Stock, ANSI Z60.1 American Association of Nurseryman, Washington, D.C.
2. Miller, F.D. and D. Neely. 1993. The effect of trenching on growth and plant health of selected species of shade trees. J. Arboric. 19: 226-229.
3. Morell, J.D. 1984. Parkway tree augering specifications. J. Arboric. 10: 129-132.

Wauconda Bank Project

Laurence R. Hall

The Wauconda National Bank is the site of a tree preservation project that our company, "The Care of Trees," has been involved with for six years, or since 1990. Prior to this, our company had performed tree care for the bank on an irregular basis. When their budget allowed, we would prune and occasionally fertilize groups of oaks and hickories that were randomly located on this two to three acre property. These are native trees, grown in what is said to be a former farm cattle pasture.

The original bank building on this site was constructed in 1976. As the bank's assets and client base grew, it became obvious that more office space was needed.Their Board of Directors decided that extending the existing building to the north, or front, of the property to accomodate an additional 9,000 square feet of office space was the expedient thing to do. This meant that construction activities would take place in and around twelve existing trees ranging in size from a 16 inch diameter hickory to a 36 inch diameter burr oak (Fig. 1).

In the spring of 1990 the Board decided to begin this project (Fig. 2). Our company was asked to cooperate with the architect, Don Erickson, to

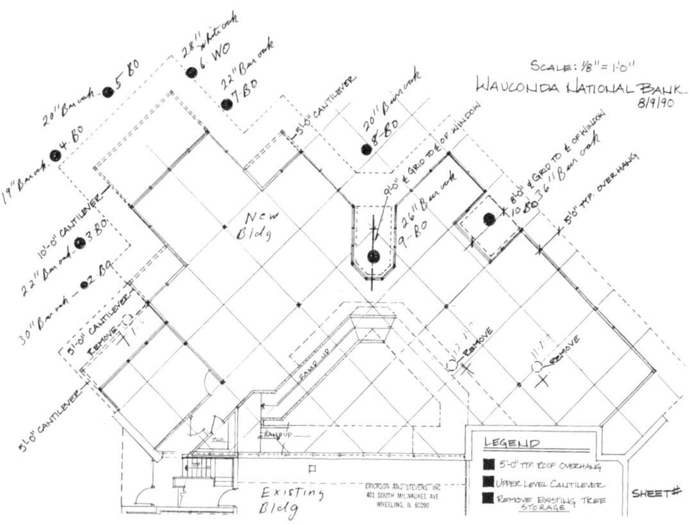

Figure 1. The architect's plan for the proposed addition to the front of the bank building showing location of existing trees.

Hendricksen, The Care of Trees, 2371 S. Foster Ave., Wheeling, IL 60090

Figure 2. A photo taken prior to construction of the bank addition. The bales of straw are placed where support pillars will be located.

develop tree preservation procedures. We presented a tree assessment list to the bank expressing our opinions as to the survival potential of each tree if all of our rules are followed. We did not give a single tree a survival rate greater than sixty percent. Four trees were in our sixty percent class, two trees were in our fifty percent class, and three trees were placed in our forty percent class. There were three trees marked for removal as we saw no chance for their survival. We stressed that the major ground work must be done on frozen ground covered with a six to eight inch layer of wood chips.

The building was to be supported by a series of concrete piers upon which steel I-beams were placed in horizontal fashion. These piers were placed in locations determined by engineers and the architect to be critical to support the new structure. We worked closely with the architect in setting these piers as far away from the trees as could reasonably be done and still have strategically placed load bearing sites to support the building.

The drilling was accomplished by a large truck-mounted earth auger. This machine was not allowed on site until we had frozen ground and had installed a six inch layer of wood chips to minimize soil compaction. The machine operator was advised of our concern for the trees and he seemed to be unusually sensitive to our cause. The initial holes drilled were twenty inches in diameter. When this auger had reached a depth of four feet below grade, a unique flaring device was used. This tool flared out below the four foot level to create an inverted cone with an ultimate diameter of eight feet. This was at ten feet below grade.

Concrete was poured into these holes which had steel reinforcement rods placed into the piers. Steel I beams were then placed upon these piers to form the base of the new building addition.

Our next trick was to build plywood forms so the concrete flooring could be poured. Our scheme was to flood the proposed building area during a cold spell and create a frozen pond. Plastic pads were to be placed upon this ice rink and concrete would then be poured onto this devised system. However, in February of 1991 we did not get an adequate cold spell to cause this hoped for solid frozen pond. We then decided to fill the area with pea gravel, place our pads and then pour concrete over this. After the concrete cured, we hired some small people to crawl beneath the concrete floor and remove most of the pea gravel.

In May of 1991 the expected truck loads of steel I beams began to arrive and assembling of the super structure began. A giant crane parked on the driveways hoisted these beams to their positions on the piers. The vertical columns were put in place and the exterior of the building was completed in late summer of 1991.

We requested that downspouts should be placed on the roof. These were set in the middle of the new addition and fed into a system of horizontal four inch diameter plastic tubes in the soil. This tubes were perforated with small holes to evenly disperse the natural rainfall over the tree root zones beneath the building. A water spigot is in place on the roof to artificially supply water in case of a drought.

A moisture meter is available to monitor soil moisture beneath the bank whenever there is a need to do so. A system of three electrodes encased in a gypsum cast has been strategically placed in the soil beneath the building. The lead wires for this system come to a single location with each wire identified so we can determine moisture levels at any or at all spots as we see fit to do so. These gypsum blocks deteriorate and are replaced every three years as per the manufacturers specifications.

Our maintenance activities include an annual fertilization which is a surface application of a 26–3–13 soluble fertilizer. This is sprayed beneath the building as well as on the wood chip mulch area which covers most of the drip line area of these trees. We prune out the deadwood from the trees every other year. When rainfall is inadequate we advise the bank to apply water on an as needed basis.

The root zones of these trees, at least to the drip line, is considered an inviolate area into which other plants are not allowed. The wood chips are topdressed every two years.

As of this writing (June 1995) the trees are doing very well. We hope this continues for years to come. This was a very interesting project from the very beginning. The bank has suggested that building costs were about forty percent more than conventional costs would have been to add 9,000 square feet of office space. The bank should be commended for their willingness to wish to save these large and beautiful trees.

The key to the success we have had to date in our tree preservation efforts on this site must be attributed to commitment by the bank to support our various recommendations. It is well known in the tree care industry that building contractors are often unwilling to listen to arborists. In this case they were told up front that the arborists could shut the job down if the trees were jeopardized. The banks support was unswerving and that made a difference.

PART II

Building and Infrastructure Damage

Infrastructure Repair Costs Associated with Street Trees in 15 Cities

E. Gregory McPherson and Paula J. Peper

Street trees are an important component of the "green infrastructure" in cities but damage caused by roots to sidewalks, curbs and gutters, and sewers is a multimillion dollar problem. To determine the magnitude of this problem, municipal foresters were surveyed in 15 cities. Total annual concrete and sewer repair costs attributed to tree damage averaged $4.28 per street tree and ranged from $0.18 to $13.65 per tree. On average, repair costs are equivalent to 25% of the annual tree program expenditures. Sidewalk repair costs are the single largest expense in all cities, averaging $3.01 per tree. Annual curb and gutter and sewer repair costs averaged $1.14 and $1.66 per tree, respectively. Damage is highly variable among cities and tends to be most severe in older areas of cities with deteriorating infrastructure and large trees. Mitigation measures applied by tree managers are discussed.

According to Kielbaso and Cotrone (9) there are an estimated 55 to 65 million existing street trees in the United States, with open spaces for planting of 65 to 75 million more. Street trees are an important component of the "green infrastructure" in cities. They provide a myriad of aesthetic, social, environmental, and economic benefits (7). In Chicago, McPherson (10) estimated that the planting of 50,000 street trees and maintaining them for 30 years would cost $8.4 million, while the benefits conferred by the trees would be $23.5 million, or $303 per tree planted. However, in many cities, damage to sidewalks, curbs, and sewer lines from tree roots is a multi-million dollar problem. Controlling these costs poses a formidable challenge to municipal foresters. The need to reduce these costs is especially great in times of dwindling municipal budgets for tree care because money saved on infrastructure repair can be well spent on other tree related activities.

To better understand the extent and magnitude of the infrastructure repair problem, we collected information from a small sample of cities regarding costs of repairing damage caused by street trees to sidewalks, curbs and gutters, and sewer or water lines. The survey addresses several questions. Are infrastructure repair costs large relative to other program expenditures? If these costs are large, what data are needed to better understand how these costs vary for trees of different species, age, and

---

E. Gregory McPherson, Research Forester and Project Leader, Paula J. Peper, Research Associate, USDA Forest Service, Pacific Southwest Research Station , Western Center for Urban Forest Research and Education. c/o Dept. of Environmental Horticulture, University of California, Davis, CA 95616

location? What are the range of repair costs across communities and do patterns exist based on typical tree lawn dimensions, street tree structure, city size, and other variables? Although sidewalk damage can result in secondary costs, such as payments resulting from physical injury claims, our analysis is limited to repair costs.

Literature Review

Infrastructure repair cost information has not been well researched and published. Only one published study compiled information from cities on infrastructure repair costs. In 1975, Hamilton and others surveyed 22 Northern California cities and reported an average annual cost of $27,000 per city for root-damaged sidewalk repair (8). More recently, two unpublished surveys compiled information on the sidewalk repair programs in California cities (6). In 1988 the League of California Cities surveyed 154 cities and in 1991 the City of Hayward Landscape Maintenance Division surveyed 14 San Francisco Bay Area cities. Eighty-two percent of all 168 cities surveyed participate in a sidewalk repair program when damage is caused by street trees in the public right-of-way. Of these participating cities, 88 percent pay 100 percent of the repair cost and 12 percent share costs with residents on a 50-50 basis. In most cases, contractors rather than the city perform repair work. In the Bay Area survey, the approximate annual cost for sidewalk repair ranged from $500 to $1000 per location.

In a sidewalk survey of residential and collector streets in San Jose, 2,274,400 square feet of sidewalk damage was found and the estimated repair cost was $14.3 million (13). The average extent of damage was 58 square feet per property and the average repair cost per property was $368. A total of 107,000 linear feet of curb and gutter damage had an estimated repair cost of $2.7 million, or 19 percent of the cost of sidewalk repair. The cost of repairing 144,300 square feet of driveway damage was estimated at $1 million. Damage was most extensive and severe in areas developed before 1964. About 68 percent of the damaged sample locations were adjacent to street trees. The species most closely associated with concrete damage were zelkova (*Zelkova serrata*), sweetgum (*Liquidamber styraciflua*), and camphor (*Cinnamomum camphora*). The report contains a number of recommendations, including implementation of an automated infrastructure management system to provide a current inventory and work management program for sidewalks, curbs, and street trees.

A sidewalk survey conducted in Oak Park, Illinois determined that on average, one of every twenty street trees was causing sidewalk heaving (defined as displacement of 1 inch or more) (14). American elm (*Ulmus americana*) and Norway maple (*Acer platanoides*) accounted for 58 and 7 percent of the all heaving incidents, respectively. Fifty-nine percent of all heaves were by overmature trees (30+ inch dbh) and 17 percent were by mature trees (20 - 29 inches dbh). Part of the sidewalk repair problem in Oak

Park was attributed to past planting practices. Although planting strip widths are spacious throughout Oak Park, years ago trees were planted within three feet of the sidewalk. Flaring root crowns of mature trees are impacting the sidewalks. Now trees are planted in the center of the planting strip to maximize distance between trees and adjacent sidewalks and curbs.

Research by Drs. Philip A. Barker and J. Alan Wagar has helped define the geometry of planting strips (2,3). They examined causal relations between trees, planting strip geometry, and soils (16). In the latter study, 763 tree sites were surveyed and regression analysis applied. Tree diameter and species were associated with over 80 percent of the sidewalk damage that was accounted for in the analysis. Planting strip width and soil texture accounted for a relatively small amount of the total variation in the regression for sidewalk damage, as well as for curb damage. Not surprisingly, the severity of sidewalk and curb damage increased with tree diameter and decreased as planting strip width increased. Damage to sidewalks exceeded damage to curbs, perhaps due to thicker and deeper concrete used in curb construction. The analysis accounted for only 38 percent of the total variation in sidewalk damage. The influence of these factors on tree root growth, infrastructure damage, and root control technologies has been the focus of much recent research and debate (17).

Results from a 1988 survey of city and county tree managers in California indicate that street tree roots are a serious problem in most of the state's cities (4). When asked to list undesirable characteristics of trees that influence selection, root damage was cited 320 times. The second most common reply was aphids at 123. Damage by roots to sewers was cited less frequently than was damage to sidewalks and curbs. Root damage was the most frequently cited reason for discontinuing use of a tree species. Twenty percent of all city and county trees that were removed in 1987 were reported to be damaging sidewalks or other structures. In the 1992 follow-up survey, planting tree species that are thought to be less likely to cause root damage was reported as the most widely used and effective method to reduce root damage (4). Root barriers are widely used, but only 25 percent of the respondents believe that they are effective. Root pruning is another commonly practiced control method whose effectiveness is not widely accepted. A frequent concern is the tendency of some species to become prone to failure after root pruning. Eliminating tree lawns by constructing sidewalks adjacent to streets and realigning sidewalks are methods used to reduce sidewalk damage from tree roots by the majority of respondents, and believed to be more effective than root barriers and root pruning (5).

In summary, studies of sidewalk and curb damage due to trees indicate that many variables are responsible for damage and their relative importance depends on conditions specific to each site. In general, the following factors may be involved: tree size, species, and proximity to concrete, planting strip width, age of concrete, soil conditions, horticultural practices (e.g., root and crown pruning, fertilization, irrigation), and other environmental conditions (e.g., vehicular and pedestrian traffic). Sidewalk damage appears to be more ubiquitous and costly to repair than damage to curbs and gutters.

We could locate only one study on tree roots and sewers. Rolf and Stal (12) note that water in and outside sewer pipes creates an environment conducive to root growth and that the "root tips follow the pipe and penetrate where there is a weakness" (p. 329). Condensation on the outside of clay pipes attracts opportunistic roots. Old concrete and clay sewer pipes without rubber gaskets in the joints are most prone to invasion by tree roots, while pipes made of PVC plastic and fiberglass are resistant if properly constructed. In older parts of many U.S. cities, clay pipe runs from buildings to sewer mains. Problems occur at the coupling of the two lines where deterioration is most common.

Once roots enter pipes, they can be cut off mechanically in roto-rooter fashion. This provides only temporary control as root regrowth can be rapid, in some cases requiring annual treatments. Chemicals can be flushed into the sewer to retard root growth, but this practice is not permitted in certain areas due to adverse impacts on water quality and plumbing systems. Relining involves using water pressure to place a rigid resin liner where the pipe is broken. A similar concept is pipe sliplining, wherein a plastic pipe is drawn or pushed through the defective pipe and ultimately joined to the original pipe with rubber rings and sockets (12). Finally, broken pipes can be replaced with new pipes. Repair costs typically range from $1,000 for excavation and replacement to $100 for rodding and roto-rooting (10).

Annual tree root removal and sewer repair costs were studied at three sites in Malmo, Sweden (12). Annual sewer maintenance costs ranged from $11 to $186 /meter ($36 to $610 /ft) of pipe. Treatment at one site, where an old concrete sewer line was penetrated by willow tree roots, included tree removal and replacement ($105 /m^2 or $1,130 /ft^2), as well as relining with acid-resistant polyester fiber impregnated with resin ($200 / m or $656 /ft). At another site relining was estimated to cost $170 / m ($558 / ft).

There are costs associated with tree impacts on other elements of urban infrastructure (e.g., street damage, invasion of water lines, blocking of street lights, and signs, leaf litter and storm debris removal). We omit these costs from this analysis while recognizing that they can be substantial in certain situations. For example, due to the nature of water main and service line construction, root damage is usually negligible. However, annual water line repair costs due to tree-related damage were reported to us to average $0.17 /tree in San Jose and $0.10 /tree in Sacramento.

Methods

We surveyed municipal foresters in 15 cities between 1991 and 1994 to gather information on tree management costs. During 1991 and 1992 we administered a questionnaire in person to arborists in Atlanta, Boston,

Dallas, Denver, Minneapolis, Rock Valley, Iowa, and Washington, D.C. These cities were 7 of 12 selected for another study that modeled benefits and costs of community tree planting. Infrastructure repair cost information was unavailable from the other 5 cities. In 1992 the questionnaire was modified and administered in person and by mail to city arborists in Chicago and Oak Park, Illinois; Hayward, Modesto, Sacramento, San Jose, and Santa Maria, California; and Vancouver, British Columbia. These cities were selected because we knew that they had information concerning street trees and infrastructure repair costs. Our sample of cities encompasses a broad geographic range, but is not representative of any particular population of cities or regions (Figure 1).

In the first questionnaire respondents were asked to estimate average annual repair costs for sidewalks, curb and gutter, and sewer lines attributed to damage caused by tree roots. Information came from a variety of sources and in different forms. In most cases, cost data were obtained from city departments of forestry, engineering, public works, transportation, sanitation, and water. Attributing the proportion of total sidewalk and curb and gutter repair costs due to tree roots was problematic. Because concrete repair is often associated with major street construction projects and these projects generally occur where infrastructure damage is most severe, it is difficult to discern the exact cause of damage. For example, concrete can be damaged by impacts from vehicles and other heavy objects, exfoliation caused by de-icing salts and extreme temperatures, and expansion and contraction of clay soil. Cities with recent sidewalk damage surveys provided annual costs for tree-related damage directly. More often, respondents gave total annual repair costs or the number of repairs made per year and average cost per repair. Then they estimated the percentage of total repair costs due to trees based on personal observation and limited field data. The accuracy of these estimates are unknown and add to the uncertainty associated with the results we present. We note when estimates of concrete repair costs were not broken out into separate costs for sidewalks and curbs and gutters,

Figure 1. Location of 15 cities surveyed for the study.

and list these data under sidewalks because repair costs were typically greater for sidewalks than for curbs and gutters.

Based on this experience, the second questionnaire asked respondents to list the years for which average annual repair costs were provided and note the estimated percentage due to trees if tree-related costs were not directly reflected in the data provided. Additional background information on the street tree population and the tree management program was requested as well. Respondents estimated the average diameter at breast height of their street tree populations since one might anticipate greater repair costs in cities with many large, old trees. Similarly, the average width of continuous planting strips was estimated, recognizing that this varies throughout the city and that "tree pits" surrounded by concrete are often common in urban core areas. Values that we present should be viewed as rough indicators of average tree sizes and planting strip widths because precise quantitative data were lacking for most cities. Because of differences in the two survey instruments, the data are not completely uniform, despite recent attempts to fill in missing information.

Estimates of the total number of street trees and total annual street tree program expenditures were obtained to analyze repair costs on a per tree and per program budget basis. We calculate the average annual cost per street tree as a basis for comparing costs across cities by dividing tree-related repair costs by the street tree population. As an indication of the relative magnitude of infrastructure repair costs, we present total annual repair costs as a percentage of the total annual tree program budget for cities that provided budget information. It should be recognized that municipal forestry budgets are not usually charged the full amount of these costs. In cities such as Chicago, sidewalk repair costs are split between the residents and the city. City funds are administered by the Transportation Department, not Forestry. In San Jose and most cities surveyed, residents are assessed the entire cost of sidewalk repair. Ultimately, citizens pay one way or another for the damage tree roots cause and it is instructive to see how this amount compares with money spent planting and caring for street trees. Because of the preliminary nature of this work and the relatively small size of the sample, we simply present estimated means and standard deviations.

In four cities (San Jose, Modesto, Sacramento, and Vancouver, B.C.) the questionnaire was followed-up with open-ended questions concerning actions being taken to reduce conflicts between trees, concrete, and sewers. Observations made by municipal foresters in each of these cities are discussed.

Results

Our sample contained cities with a wide range of human and street tree populations (Table 1). The number of street trees per capita in our sample ranged from 0.08 in Boston to 1.11 in Rock Valley, Iowa, with a mean of 0.32.

This mean of 0.32 street trees per capita compares favorably with a mean of 0.37 found for 22 street tree populations (11), but is substantially less than the mean of 0.63 obtained from a recent survey of 419 municipalities (15). The latter discrepancy is probably due to inclusion of park and other public trees in their survey. Annual tree program budgets ranged from $4,000 in Rock Valley to $11.3 million in Chicago. Our sample mean of $2.56 million is substantially greater than the mean of $279,307 reported in the 1994 survey of 419 cities (5). In the following sections we report sidewalk repair costs for 15 cities, curb and gutter repair costs for 5 cities, and sewer repair costs for 8 cities.

Sidewalk repair costs. Annual sidewalk repair costs due to tree-related damage are estimated to be under $1 per tree in four cities (Minneapolis, Rock Valley, Boston, and Denver) and $6 or greater per tree in Modesto, Vancouver, B.C., and Hayward (Table 2). Costs for Modesto and Hayward are inflated because curb and gutter repair costs are not broken out from sidewalk repair costs. The mean repair cost is $3.01 per tree and the standard deviation of $2.50 reflects the large variability in repair costs among cities.

Table 1. Background information on the sample cities.

City	Region[a]	Pop. (1,000s)	Area sq miles	Pop./ sq mile	Program Budget ($)	Tree Pop.	Trees/ capita	Plt. Strip Width (ft)	Avg. Dbh (in)
Atlanta, GA	South	437	132	3.3	800,000	70,000	0.16	3.5	21.0
Boston, MA	GL-NE	520	46	11.3	865,000	44,000	0.08	3.5	15.0
Chicago, IL	GL-NE	2,780	237	11.7	11,330,000	459,000	0.17	4.0	16.0
Dallas, TX[b]	Cent. TX	1,000	331	3.0		240,000	0.24	4.0	10.0
Denver, CO[b]	Mt	468	155	3.0		130,000	0.28		
Hayward, CA	CA Coast	122	61	2.0	1,250,000	30,000	0.25	3.5	24.0
Minneapolis, MN	N. Tier	375	59	6.4	5,800,000	143,000	0.38	5.0	25.0
Modesto, CA	Semi-Arid	182	34	5.4	2,068,000	75,000	0.41	4.0	12.0
Oak Park, IL	GL-NE	54	5	12.0	887,000	19,000	0.35	7.0	18.0
Rock Valley, IO	GL-NE	3	1	3.0	4,000	3,340	1.11	8.0	15.0
Sacramento, CA	Semi-Arid	385	98	3.9	3,100,000	115,000	0.30	5.0	24.0
San Jose, CA	CA Coast	838	173	4.8	1,700,000	250,000	0.30	3.5	13.0
Santa Maria, CA	CA-Coast	78	18	4.3	245,000	25,000	0.32	6.0	14.0
Vancouver, BC	Pac. NW	509	44	11.6	2,205,975	102,510	0.20	5.0	13.5
Washington, DC	NE	623	69	9.0	3,000,000	110,000	0.18	4.0	15.0
Mean (SD)		558 (655)	98 (90)	6.3 (3.6)	2,558,075 (2,927,071)	121,057 (115,036)	0.32 (0.23)	4.7 (1.4)	16.8 (4.7)

[a] Regions represent distinct climates and population centers (1).
[b] Blanks indicate data were unavailable.

Sidewalk repair costs are equivalent to 40 percent of San Jose's annual tree program budget, although they are estimated to be only $2.73 per tree. In six other cities annual sidewalk repair costs are estimated to be equal to 20 percent or more of their respective program budgets (Atlanta, Hayward, Modesto [includes curb and gutter], Rock Valley, Vancouver, and Washington, D.C.) (Table 3). In Boston, Minneapolis, and Oak Park, Illinois, sidewalk repair costs are less than 5 percent of annual tree programs expenditures.

Curb and gutter repair costs. Annual repair costs for curb and gutter damage due to trees range from $0.18 per tree in Oak Park to $3.51 per tree in Sacramento (Table 2). The mean for our sample of 5 cities is $1.14. The mean cost for curb and gutter repair is 38 percent of the mean sidewalk repair cost. Although repair costs are less for curbs and gutters than sidewalks in all 5 cities, differences are surprisingly small in Denver ($0.87 versus $0.96) and Sacramento ($3.51 versus $3.97).

In Sacramento, curb and gutter repair costs are equal to 13 percent of the annual tree program budget (Table 3). Values for other cities range from 0.4

Table 2. Reported annual costs for sidewalk and curb and gutter repairs.

City	Sidewalk			Curb and Gutter			Total Concrete	
	Total Repair Cost ($)	Tree-Related (%)	Cost/ Tree ($)	Total Repair Cost ($)	Tree-Related (%)	Cost/ Tree ($)	Total Repair Cost ($)[a]	Cost/ Tree ($)
Atlanta, GA	161,000	100	2.30				161,000	2.30
Boston, MA	15,300	100	0.35				15,300	0.35
Chicago, IL	3,000,000	33	2.16				990,000	2.16
Dallas, TX	3,000,000	20	2.50				600,000	2.50
Denver, CO	125,000	100	0.96	750,000	15	0.87	237,500	1.83
Hayward, CA[b]	244,000	100	8.13				244,000	8.13
Minneapolis, MN	25,176	100	0.18				25,176	0.18
Modesto, CA[b]	450,000	100	6.00				450,000	6.00
Oak Park, IL	125,000	20	1.32	17,500	20	0.18	28,500	1.50
Rock Valley, IO	800	100	0.24				800	0.24
Sacramento, CA	608,115	75	3.97	673,177	60	3.51	859,992	7.48
San Jose, CA	1,003,904	68	2.73	150,584	68	0.41	785,052	3.14
Santa Maria, CA	36,500	100	1.46				36,500	1.46
Vancouver, BC	669,750	100	6.53	75,000	100	0.73	744,750	7.27
Washington, DC	700,000	100	6.36				700,000	6.36
Mean (SD)	677,636 (957418)		3.01 (2.50)	333,252 (312,726)		1.14 (1.21)	391,905 (344,885)	3.39 (2.74)

[a] Totals are for tree-related damage to sidewalks and curbs and gutters.
[b] Sidewalk and curb and gutter repair costs are not broken out.

percent in Oak Park to 3.4 percent in Vancouver B.C.

Total annual concrete repair costs, calculated as the sum of sidewalk and curb and gutter costs, average $3.39 per tree (Table 2). Cities spending the most per tree to repair concrete damage are Hayward ($8.13), Sacramento ($7.48), Vancouver ($7.27), and Washington, D.C. ($6.36).

Sewer repair costs. Sewer repair costs range from $0.11 per tree in Boston to $6.39 per tree in Vancouver, B.C. (Table 4). Other cities with costs over $1 per tree are Atlanta ($1.71) and Sacramento ($3.40). The mean annual repair cost is $1.66 per tree for 8 cities. Vancouver's relatively high costs are due the large number of clay and concrete lines that are over 100 years old. Also, the lines were placed under the middle of the planting strip and trees were planted directly above the lines. Today the city has a deteriorating sewer system with large, old trees placed in perfect position to invade the lines.

Expenditures associated with repairing tree damaged sewer lines are

Table 3. Total annual repair costs per tree and as a percentage of total annual street tree program budget.

City	Total Costs Per Tree ($)	Percent of Total Annual Program Budget			
		Sidewalk	Curb & Gutter	Sewer	Total
Atlanta, GA	4.01	20.1		15.0	35.1
Boston, MA	0.46	1.8		0.6	2.3
Chicago, IL	2.92	8.7		3.1	11.8
Dallas	2.50				
Denver, CO	1.83				
Hayward, CA[a]	8.13	19.5			19.5
Minneapolis, MN	0.18	0.4			0.4
Modesto, CA[a]	6.00	21.8			21.8
Oak Park, IL	2.01	2.8	0.4	1.1	4.3
Rock Valley, IO	0.51	20.0		22.5	42.5
Sacramento, CA	10.88	14.7	13.0	12.6	40.4
San Jose, CA	3.26	40.2	6.0	1.8	47.9
Santa Maria, CA	1.46	14.9			14.9
Vancouver, BC	13.65	30.4	3.4	29.7	63.4
Washington, DC	6.36	23.3			23.3
Mean (SD)	4.28 (3.87)	16.8 (11.1)	5.7 (4.7)	10.8 (10.3)	25.2 (18.6)

[a] Sidewalk and curb and gutter repair costs are not broken out.

Table 4. Reported annual costs for sewer repairs.

City	Total Repair Cost ($)	Tree-Related (%)	Cost/ Tree ($)
Atlanta, GA	120,000	100	1.71
Boston, MA	5,000	100	0.11
Chicago, IL	350,000	100	0.76
Oak Park, IL	9,600	100	0.51
Rock Valley, IO	900	100	0.27
Sacramento, CA	2,059,512	19	3.40
San Jose, CA	100,000	30	0.12
Vancouver, BC	727,500	90	6.39
Mean (SD)	421,564 (661,072)		1.66 (2.07)

equivalent to 13 percent or more of the annual tree program budgets in Sacramento (13%), Atlanta (15%), Rock Valley (20%), and Vancouver (30%) (Table 3). Percentages for the remaining four cities are 3.1 percent or less.

Total repair costs. On a per tree basis, total annual concrete and sewer repair costs attributed to tree-related damage average $4.28 (standard deviation of $3.87) and range from $0.18 in Minneapolis to $13.65 in Vancouver, B.C. (Figure 2 and Table 3). Scatter plots showed no patterns of relation between cost per tree and independent variables such as city population, population density, average planting strip width, or average tree dbh. This finding is not surprising given the site specific nature of infrastructure damage and the city-wide nature of the predictor variables selected. It may be possible to more accurately predict repair costs using similar variables applied to specific streets or neighborhoods within a city, but to our knowledge this has not been accomplished.

Total concrete and sewer repair costs average 25 percent of total tree program budgets (19% deviation) and range from 0.4 percent in Minneapolis to 63 percent in Vancouver, B.C. (Figure 3 and Table 3). Rock Valley, Iowa and San Jose had relatively low costs per tree but relatively high costs as a percentage of program budget. This fact is primarily due to their relatively small program budgets but relatively large street tree populations.

Results of Survey Follow-Up

Tree species belonging to genera such as *Liquidambar*, *Fraxinus*, *Zelkova*, *Gleditsia*, and *Prunus* topped the lists of those most often associated with concrete damage, with *Liquidambar* causing sidewalk, curb and gutter upheaval within 15 to 20 years of planting (about 10-12 inches dbh).

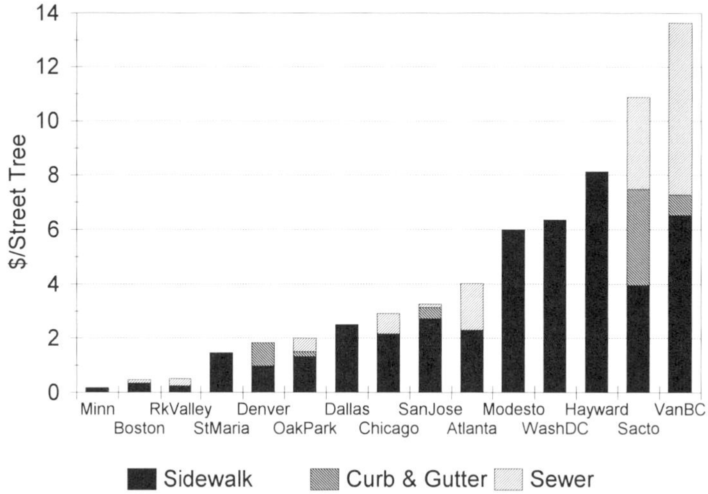

Figure 2. Annual infrastructure repair costs per street tree.

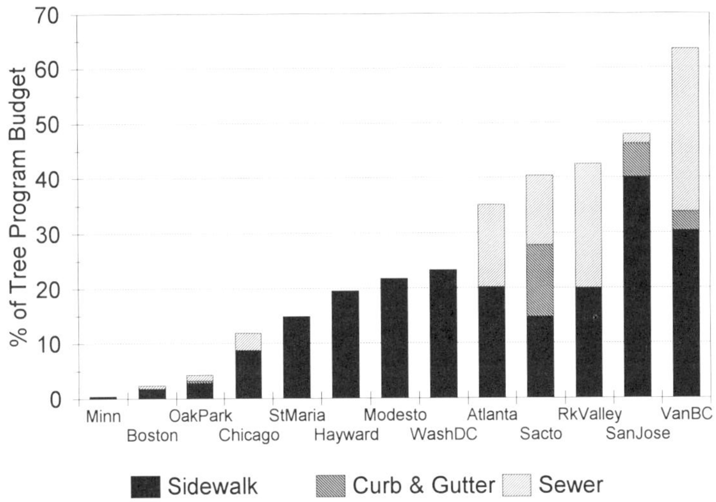

Figure 3. Annual infrastructure repair costs as percentages of annual tree program budgets.

However, the species listed by tree managers also represented the dominant species planted in each of their cities. All managers agreed that damage is less species specific and more site specific, noting that damage tended to be most severe with plantings of trees, aged thirty and older, growing in narrow planting strips. At sites where trees repeatedly uplifted sidewalks, the concrete replacement cycle ranged from five to ten years, whether or not root cutting was part of the replacement procedure. The Sacramento Tree Services Division has adopted a policy of removing "repeat offender" trees if the concrete requires replacement more than once every eight years at sites where sidewalk relocation is not feasible. Removal still remains a last resort option in Sacramento, as it does in the other three cities contacted.

Measures the cities specifically apply to avoid trip and fall accidents and the possibility of litigation due to concrete upheaval include yearly visual inspections, enforcing policies of homeowner repair of damaged sidewalks, curbs and gutters, and grinding pavement uplifts as a short term remedy until replacement is possible.

Respondents' repair methods consist of variations on the general theme of removing old concrete, root-pruning or cutting, and pouring new concrete. San Jose routinely cuts roots to an 18-inch depth during sidewalk removal and replacement. Homeowners in Sacramento, who are responsible for all concrete replacement, are required to contact the City Tree Services Division to inspect the site once concrete is removed so a determination can be made as to the cause of the damage. If roots are the cause, the city prunes and removes them before the sidewalk is replaced. The inspector may recommend that the sidewalk be curbed or bowed out around the tree or that the width of the sidewalk be reduced in order to increase planting space and reduce the chance of damage recurrence. In Vancouver, current replacement methods include relocating sidewalks where possible and using asphalt or crushed rock sidewalk inserts in lieu of concrete where repeated problems occur.

Although respondents listed species selection and diversity as key components in avoiding concrete damage (particularly where planting space cannot be increased), all are considering, and in some cases applying, methods developed to deter damage and lengthen the concrete replacement cycle. When fiscally possible, replacement sidewalks in Modesto are poured with and 8-inch deep apron on the tree side to act as a root barrier. Chemical barriers are also being used as an experimental method for deflecting potentially invasive roots. Other cities are using a variety of physical root barrier products on an experimental basis and San Jose is proposing the use of trenching and copper screening. Two of the cities are tackling the root damage problem from a engineering perspective. Sacramento has experimented with a foam additive for concrete designed to produce a more elastic sidewalk, one which would bend rather than break and lift, as roots beneath grow and expand. In a similar vein, Vancouver is experimenting with sidewalks engineered with an air gap left between the bottom of the walk and the soil surface. The use of trenches backfilled with a growing medium that will direct roots away from concrete is being considered in conjunction with these air gaps, as well.

All four cities share a common desire to retain existing trees while designing and planning a future urban forest that is less damaging to the infrastructure, therefore less costly to maintain. In line with results published in Bernhardt and Swiecki's surveys (4,5), several of the tree managers stated that root damage to sidewalks, curbs and gutters is far more pervasive and costly than the damage to sewers. Moreover, the damage to concrete structures is directly attributed to trees while damage to sewer lines is not. In addressing damage to sewer lines, our respondents refer to tree roots as opportunists taking advantage of the fertile, moist environment provided by deteriorating clay and concrete pipe. They consistently report root intrusion occurring only where there are pre-existing leaks in the sewer lines. These managers also confirm Rolf and Stal's (12) report that clay and concrete pipes "sweat" more than PVC plastic types, and roots are tending to run along pipe exteriors, invading pipe interiors at leaking joints or other breaks in the lines. As these invasive roots expand, the pipe joints often rupture. None of the respondents reports using chemicals to purge invasive roots from lines. Root cutting remains the preferential method for removing blockages. PVC pipe and couplings are used for repairs whenever possible and plastic-to clay rubber couplings used to connect new pipe to old systems. This coupling can be a weak link in the newly replaced section since several managers commented on problems with slippage as the pipe expands and contracts, potentially creating new leaks that will attract roots. To date, however, this repair method remains the financially feasible option for cities with thousands of linear feet of old clay and concrete sewer systems in need of replacement.

Overall, responses to our questions indicate that the tree managers are testing a variety of methods to reduce root damage to the infrastructure, all focusing on methods to promote retention of trees. None could list any one method as the key to damage reduction. They tend to view the problem as site-related, involving interrelationships between tree species, tree growth and spatial requirements, and existing site limitations.

Summary and Conclusions

Costs for repairing the infrastructure damaged by street trees are substantial. Total annual costs are estimated to average $4.28 per tree for our sample of 15 cities. On average, this cost amounted to 25 percent of the sample cities' total annual tree program budgets. A national survey of 419 cities found that the mean annual expenditure per public-owned tree is only $4.65 (15). Our estimates of infrastructure repair costs are conservative because data on curb and gutter and sewer repair costs are lacking for half the cities in the sample. Also, sewer repair costs paid by residents but caused by street trees are not included. Data from sewer repair contractors are needed to better estimate the magnitude of these costs. Costs are self-reported and may reflect the bias of individuals. For example, individuals who view trees as the "enemy" may report higher costs than those who view trees as an "amenity."

Repair costs are extremely variable from city to city with no strong relations evident between costs and predictor variables such as city population, population density, average tree dbh, and average planting strip width. Within a given city, costs are highest in older areas where infrastructure is deteriorating and trees are likely to be large. Cost-effective mitigation that preserves benefits from existing tree cover while reducing repair costs is critical in these situations. In new developments, repair costs can be minimized by locating trees and infrastructure to minimize conflicts, using "tree resistant" materials, and selecting species that are most suitable for site conditions.

Many tree managers and several researchers are conducting experiments to evaluate the effectiveness of different mitigation measures, particularly related to concrete damage. This work is extremely important given the magnitude of infrastructure damage associated with street trees. Close collaboration between scientists and managers is needed to insure that experiments are designed to produce results that will be statistically sound and published for the benefit of a wider audience. Survey and field research that document the extent of concrete and sewer damage due to trees at specific sites over long periods of time are needed to better understand the nature of tree-related damage. Data required to identify the roles of different factors and to evaluate the long term cost-effectiveness of different design and mitigation measures include:

- full documentation of design details
- extent of concrete/sewer damage and repair costs
- type, location, and age of the infrastructure
- tree species, location, health, and growth over time
- environmental conditions below ground (e.g., soil type, bulk density, moisture), at the surface (e.g., planting strip width, cover type, site uses), and above ground (e.g., atmospheric conditions, management practices [pruning, irrigation, fertilization])
- repair methods, installation and on-going maintenance costs, and durability over time.

Understanding cause and effect relations between trees and infrastructure damage is difficult because of complex interactions among the many factors involved at each particular site. Developing legitimate research results that will guide managers in their efforts to reduce infrastructure repair costs will require combining the practical knowledge of managers with the scientific expertise of researchers.

Acknowledgments

Drs. Patricia Lindsey (University of California, Davis) and Robert Tate (Robert L. Tate Associates) provided helpful reviews of an earlier version of this paper. We thank the municipal tree managers for their cooperation providing the data used in this study. They include: Jay Lowery, Brian Gilbert, Polly Melton, Steve Bylina, Claude Thompson, Kassandra McLaughlin, Paul Ryan, Judity O'Conner, Jim Hermann, Ralph Sievert, Chuck Gilstrap, Mike Stankovich, Jim VanderVelde, Martin

Fitch, Mark Beaudoin, Bailey Hudson, Paul Montpellier, and Sandra Hill. Also, we thank Tom Pehrson of Urban Forestry Consultants for his assistance.

Literature Cited

1. Andersson, B., W. Carroll and M. Martin, 1986. Aggregation of U.S. population centers using climate parameters related to building energy use. J. Clim. & Appl. Meteorol. 25: 596-614.
2. Barker, P.A. 1976. Planting strips in street rights-of-way: a key public land resource. pp. 263-274. J.W. Andresen (ed.). Trees and Forests for Human Settlements. University of Toronto Press, Toronto, Ontaria, Canada.
3. Barker, P.A. and M.G. Durrant 1978. Space for trees in street rights-of-way. pp. 463-467. North America's Forest: Gateway to Opportunity, Proceedings of the Society of American Foresters and the Canadian Society of Foresters. Society of American Foresters, Bethesda, MD.
4. Bernhardt, E. and T.J. Swiecki 1988. The state of urban forestry in California: Results of the 1988 California Urban Forest Survey. California Department of Forestry and Fire Protection, Sacramento, CA.
5. Berhardt, E. and T.J. Swiecki 1994. The state of urban forestry in California-1992. California Department of Forestry and Fire Protection, Sacramento, CA.
6. City of Hayward 1993. Implementation of the 1993-94 sidewalk rehabilitation program. City of Hayward, Hayward, CA.
7. Dwyer, J., G. McPherson, H. Schroeder and R. Rowntree. 1992. Assessing the benefits and costs of the urban forest. J. Arboric. 18: 227-234.
8. Hamilton, D., W. Owen and W. Davis 1975. Street tree root problem survey. University of California Cooperative Extension, Alameda County.
9. Kielbaso, J.J. and V. Cotrone. 1990. The State of the Urban Forest. pp. 11-18. P.D. Rodbell (ed.). Make our cities safe for trees. American Forestry Association, Washington, D.C.
10. McPherson, E.G. 1994. Benefits and costs of tree planting and care in Chicago. pp. 115-134. E.G. McPherson, D.J. Nowak and R.A. Rowntree (eds.) Chicago's Urban Forest Ecosystem: Results of the Chicago Urban Forest Climate Project. Northeastern Forest Experiment Station, Radnor, PA.
11. McPherson, E.G. and R.A. Rowntree. 1989. Using structural measures to compare twenty-two U.S. street tree populations. Landscape Journal 8: 13-23.
12. Rolf, K. and O. Stal. 1994. Tree roots in sewer systems in Malmo, Sweden. J. Arboric. 20: 329-335.
13. Sealana & Associates 1994. City of San Jose sidewalk survey & analysis study, City of San Jose, Department of Streets and Parks, San Jose, CA.
14. Stankovick, M. 1990. Sidewalks and trees. Village of Oak Park, Oak Park, IL.
15. Tschantz, B.A. and P.L. Sacamano. 1994. Municipal Tree Management in the United States. International Society of Arboriculture, Savoy, IL.
16. Wagar, J.A. and P.A. Barker. 1983. Tree root damage to sidewalks and curbs. J. Arboric. 9: 177-181.
17. Watson, G.W. and D. Neely. (eds.) 1994. The Landscape Below Ground. International Society of Arboriculture, Savoy, IL.

Root Barriers Affect Root Distribution

Edward F. Gilman

There is one published report in the scientific literature of newly-planted trees' response to a physical barrier designed to deflect tree roots (3). It shows that a plastic barrier deflects roots to deeper soil layers; however after growing under the barrier, in a poorly drained soil, the roots grew quickly back to the soil surface. The barrier was most effective in the parts of the field containing well-drained soil where roots grew back to the surface more slowly. Other reports indicate that barriers can deflect roots cut on established trees and can potentially reduce damage to sidewalks (2). Approximately 25% of the nearly 200 urban forestry programs using root barriers in California report them to be effective in reducing sidewalk damage; another 25% report that they are partially effective (1). Most responding to the survey say that species selection is nearly 90% effective in reducing sidewalk damage.

The purpose of this study was to compare root growth on recently transplanted trees near and under Biobarrier® with root growth near similar trees without the barrier.

Materials and Methods

In late winter 1991, 30-cm-deep Biobarrier (REEMAY, Inc.) was installed vertically in loose, sandy, poorly-drained soil in Gainesville Florida (USDA hardiness zone 8b). Two, 5-cm-wide 30-cm-deep, parallel trenches 1.5 m apart were dug with a carbide tipped trencher and a continuous piece of Biobarrier was slipped into each open trench. Soil was manually replaced in the trench after slipping the Biobarrier into the trench. This arrangement was installed in two sections of the field. About 3 cm of the barrier was left above the ground. The water table at the site was 30 to 45 cm below the soil surface in winter and early spring and about 1.3 m during the rest of the year.

Seven container-grown, 5-cm caliper live oak (*Quercus virginiana*) and 7 sycamore (*Platanus occidentalis*) were alternately planted 1.8 m apart centered between the parallel strips of Biobarrier. Seven trees of each species were also planted 1.8 m apart without Biobarrier. The top of the container root ball was placed even with the soil surface. An 8 cm thick layer of wood chip mulch was applied to the soil surface between the parallel strips of Biobarrier. The trunks were 75 cm from each strip of Biobarrier.

To encourage rapid growth after planting, root balls were irrigated daily for several months then every other day until the beginning of winter 1992.

Dr. Gilman is with the Department of Environmental Horticulture, University of Florida, Gainesville, FL 32611.

Irrigation was then discontinued. Fertilizer was applied every four months at a rate of 2 pounds/1000ft^2/application.

In spring 1994, a 1.2 m wide and 1.2 m deep trench was dug with a backhoe 15 cm away from and parallel to the outside of the Biobarrier on one side of a group of six trees (3 of each species). The trench was dug this wide to simulate the width of a 1.2 m (4 ft) wide sidewalk. As the trench was deepened with the backhoe, a sharpened square shovel was used to make clean, vertical cuts in the soil at the edge of the trench so roots would be cut cleanly. A trench with the same dimensions was dug the same distance from six control trees.

One wall of the trench was 0.91 m away from the trunks (15 cm away from the Biobarrier) and the other was 2.13 m from the trunk. In essence, if a sidewalk had been poured at the site, one vertical wall would be at each edge of the sidewalk.

A 1.8 m long x 60 cm deep piece of transparent plastic was fixed to the face of each wall centered at the tree trunk. All roots greater than 3 cm diameter were located on the plastic by tracing the outside edge of the root. Roots in each soil depth class (0-15 cm, 15-30 cm, 30-45 cm and 45-60 cm) were counted. Data were analyzed with Analysis of Variance in a randomized complete block design. The percentage of total root number on the tree that was located in each root depth class was also calculated.

Results and Discussion

More 3 - 5 cm diameter roots were found on both species in the Biobarrier treatment than in the control. There were no other differences in root number between treatments for any other root diameter class. Perhaps the trifluralin in Biobarrier caused an increase in root branching. No roots of either species went through the Biobarrier. Wagar and Barker (2) also found that no birch (*Betula papyrifera*) roots penetrated the Biobarrier, but some *Populus trichocarpa* x *deltoides* and *Populus trichocarpa* roots did.

Most roots on both species without a barrier were located in the top 30 cm of soil (Table1). Root number decreased with soil depth. Root distribution was different for Biobarrier trees. The number of roots 15 cm from Biobarrier increased with soil depth with most roots 30 to 45 cm below the soil surface. There appeared to be no question that the barrier pushed roots into deeper soil. Despite the increase in the number of deep roots for the Biobarrier treatment, many roots 15 cm away from the barrier were still located in the top 15 cm of soil. Although the data do not show it, many of these roots were in the top 8 cm. This resulted from roots that grew nearly straight up toward the soil surface once they grew under the barrier. These roots often were only several cm below the soil surface once they turned parallel to the soil surface. Wagar (3) also reported that roots in poorly drained soil quickly returned to the soil surface after growing under a physical barrier.

Table 1. Root number in various depth classes 91 cm away from tree[1].

Depth from soil line (cm)	Control (no barrier) Root diameter class		Biobarrier Root diameter class	
	< 10 mm	> 10 mm	< 10 mm	>10 mm
Live oak				
0 - 15	30	4	22	4
15 - 30	22	4	20	6
30 - 45	6	0	40	6
45 - 60	9	0	26	0
Sycamore				
0 - 15	28	11	14	3
15 - 30	18	4	19	8
30 - 45	11	3	38	5
45 - 60	6	3	25	3

1. This was 15 cm outside the Biobarrier.

Table 2. Percent of roots in depth classes for live oak planted 0.7 meters from Biobarrier or with no barrier.

Depth from soil line (cm)	Distance from trunk (m) 0.91		2.13	
	Biobarrier	Control	Biobarrier	Control
0 - 15	21	45	53	88
15 - 30	21	35	22	12
30 - 45	37	8	7	0
45 - 60	21	12	18	0

The effect of the root barrier on root distribution can be presented in another manner (Table 2). Eighty percent of roots 0.91 m from the trunk on live oak without a barrier were in the top 30 cm of soil; whereas, only 42% of roots were in the top 30 cm for trees with the root barrier. On the trench wall another 1.2 m away from trees with the barrier, 75% of live oak roots were in the top 30 cm. More than half of the roots were in the top 15 cm and some of these were just under the surface of the soil.

Summary

Biobarrier pushed some roots deeper in the soil but many of these made their way back to the surface by the time they had grown 1.2 m away from the barrier. Some roots grew nearly straight up after growing under the barrier in what appeared to be the original trench dug to initially install the barrier.

Acknowledgment

Thanks to REEMAY Corporation, Old Hickory, Tenn. for funding part of this study.

Literature Cited

1. Bernhardt, E. and T.J. Swiecki. 1993. The state of Urban forestry in California - 1992. California Department of Forestry and Fire Protection. pg 19.
2. Wagar, A.J. 1985. Reducing surface rooting of trees with control planters and wells. J. Arboric. 11(6): 165-171.
3. Wagar, A.J. and P.A. Barker. 1993. Effectiveness of three barrier materials for stopping regenerating roots of established trees. J. Arboric. 19(6): 332-338.

Tree Roots and Sewer Systems

Kaj Rolf, Örjan Stål and Håkan Schroeder

An expensive problem which can lead to serious conflicts of interest between arborists and sewer companies is tree roots entering sewer pipes. A substantial amount of money is spent each year on different measures to remove the roots after they have grown into pipes. The most frequently used methods to prevent root penetration are designed for conventional sewer construction. Root cutting, high pressure flushing and relining are solutions that give an immediate result but the effects are often transitory and sometimes will even increase the problem in the long term since root cutting stimulates root growth in a particular area. In a research project at the Swedish University of Agricultural Sciences the problem has been quantified and new alternative strategies are being developed. It is not possible to recommend one solution that can cure every root penetration problem. Instead it is necessary to investigate each problem area separately. In some cases it may be best to remove the trees whilst in other places they represent an important aesthetic or environmental value. An evaluation of technical and economic aspects of the options should be combined with aesthetic and cultural considerations. This review of conditions of specific importance for intrusion of roots in sewer pipes identifies the necessity of communication between all people involved in planning and maintenance of trees and sewers. The possibilities of applying different preventive or corrective measures to eliminate or reduce the problem are discussed.

Tree roots cause damage to sewer pipes and trenching near trees for the installation of underground utilities causes damage to tree roots. Thus, there are large economic investments involved when there is a conflict between trees and sewer systems. Each party involved defends his own interests and seldom is there a satisfactory solution to the problem. A constructive dialog between the parties involved may save expenses and prevent trees from being felled. Different opinions between the people involved in underground utility lines and the people with interests in the trees may arise when:

- trenching is done near existing trees
- trees are planted near or on existing underground installations
- tree roots grow into sewer systems.

Tree roots entering sewer systems is a well-known phenomena. As early as the nineteen thirties, copper wire was put into the joints on drainage pipes to prevent tree roots from growing into the pipe. A report from a survey about the extent and quantity of root penetration from 242 of Sweden's 286

The authors are with the Swedish University of Agricultural Sciences, Department of Agricultural Engineering, Section of Horticultural Engineering, Box 66, S-230 53 Alnarp, Sweden

Wastewater Works departments shows that 99% of those who have replied have had problems with root intrusion in sewers and that the problem occurs also with newer types of pipes (10).

A prerequisite of root intrusion in sewer pipes is, of course, trees near the pipe. However, no research has yet examined situations to determine precisely when and where roots penetrate sewer systems. A common method of preventing this has been to remove the trees.

Cooperation between all people involved in the building industry is an absolute necessity to ensure a green environment and to minimize the costs. This cooperation has been poor or nonexistent. Sites have been planned in detail for buildings and their technical support with roads, electricity, telephone and underground installations. The planning for green urban space has seldom been as good. Another common problem in the planning process is that numerous planners involved in urban green areas seem to have a poor knowledge of the problems different plant species can cause to technical installations.

Root Development and Risk Judgment

Generally, it may be stated that big trees and fast-growing trees cause the biggest problems. However, it is necessary to bear in mind that trees develop differently; some grow fast, some are slower, some have large root systems, some have small. To be able to chose the right tree for the right place, it is important to know how the root systems of different species develop normally and how they are affected by the soil conditions.

Cutler and Richardson (1) compiled data from 3000 trees where tree roots had caused damage to buildings. This included data from different tree species and tree sizes that had been growing around and near buildings and underground installations. Table 1 is a summary of these data. Tree species has a large influence on the risk of damaging sewer systems but tree age and where the tree is growing may be of even larger influence.

Simplified, a tree's root system can be described as what the environment allows it to be. Usually the root systems of trees can be divided into three root forms; taproot, heartroot and flatroot. Some species have a very distinct form of root system while others are a mixture of the different types. The root forms that best avoid root intrusion in pipes is the tap- and heartroot forms because these roots do not grow so wide. Roots with flatroot form tend to grow far beyond the dripline. However, it should be remembered that these forms of root systems are most significant when the tree is young and that soil conditions in an urban situation hardly permit the typical root form for a specific tree to develop.

To be able to live and develop, a tree must have a sufficient soil volume and a suitable soil environment. The condition of the soil may be the factor that has been taken least into consideration when the reasons for root intrusion in sewer pipes and drains have been discussed. When the

Table 1. Distance from the tree where there is a risk of tree roots damaging buildings and drains. Maximum tree-to-damage distance recorded and the distance from the tree within which 90 % of damage occurred [after Cutler and Richardson (1)].

Tree species		Risk of damaging buildings and drains	
Scientific name	Common name	Max. recorded	90% damage
Acer sp.	Maples	20 m	12 m
Aesculus hippocastanum	Horse Chestnut	23 m	15 m
Betula sp.	Birches	10 m	8 m
Carpinus betulus	Common Hornbeam	17 m	
Crataegus sp.	Thorns	11.5 m	8.7 m
Fagus sylvatica	Common Beech	15 m	11 m
Fraxinus excelsior	Common Ash	21 m	13 m
Malus sp, Pyrus sp	Apples & Pears	10 m	8 m
Platanus sp	Planes	15 m	10 m
Populus sp	Poplars	30 m	20 m
Prunus sp	Cherries & Plums	11 m	7.5 m
Quercus sp	Oaks	30 m	18 m
Robinia sp	False Acacia	12.4 m	10.5 m
Salix sp	Willows	40 m	18 m
Sorbus intermedia	Swedish Whitebeam	11 m	9.5 m
Sorbus aucuparia	Rowan	11 m	9.5 m
Tilia	Limes	20 m	11 m
Ulmus sp	Elms	25 m	19 m
Chamaecyparis sp	Cypresses	20 m	5 m
Pinus sp	Pines	8 m	

conditions in the planting pit are not good, tree roots may grow into the nearby pipe trench (Figure 1). In this pipe trench it is easier for the root to penetrate the backfill. Another reason for the good environment for root growth around pipes is the condensation of water outside the pipe. The root tip then follows the pipe and penetrates where there is a weakness.

Type and Condition of the Pipe

The design and type of the pipe is of utmost importance for root intrusion and the amount of root damage. Other factors of importance are the age of the pipe and the stresses on the pipe material. Different materials resist root intrusion differently. Concrete pipes without rubber gaskets in the joints, in combination with old age of the pipes, are those that can resist root intrusion the least and make up the majority of pipes with root intrusion problems. Concrete pipes with rubber gaskets in the joints have been found to have a higher resistance against root intrusion but are not fully reliable, especially not in older installations. The weakness with concrete pipes is that the durability is affected by high water flow, chemicals and changes in direction

Figure 1. An illustration of tree roots growing into a nearby pipe trench, when there is a compacted subsoil that the roots cannot penetrate.

due to bad pipe bedding. Directional changes of this kind will cause great strains on the joint and the packing material.

Pipes made of plastic (PVC) and fiberglass (GAP) have, so far, been nearly resistant to root intrusion, if we disregard carelessness during construction, and joints between these types of pipes and concrete pipes. Polythene pipes (PEH) are welded so there are no joints and no problems with root intrusion. The plastic pipes must have a well-constructed and compacted gravel bedding in order to get the stability to withstand a physical load, or there is a risk for deformation and cracking, with root intrusion as a result.

Planning Measures

Cooperation between the different professions involved is essential for good results. The proprietor, designer, local manager and contractors should together find the best solution for the area to be built. In that way all interests will be considered. This cooperation is important even when it comes to renovations and repair. Some examples of preventative measures can be:

- All underground installations should be located mainly in paved areas. If this is not possible, pipes should be laid in a pattern that interferes with the trees as little as possible. Underground installations should be confined to specific areas. Always use pipes that leak as little as possible.
- Trees should be planted as far from the pipes as possible. If the tree must stand near a pipe, use trees with less active, nonaggressive root systems.

The trees should be given a well designed planting pit.

- During construction you will have to decide whether or not a tree should be saved. If you are unsure whether the tree will survive construction work or will cause damage to underground installations, it is better to remove the tree and replant with a new one. In that case it is easier to create better conditions for both the tree and the pipe.

Construction work near existing trees. When a new pipe is to be installed in places where trees are already growing, the area must be thoroughly examined before the exact location of the pipe is decided. Find out the age, tree species and root and soil volume. To decide the distance between the pipe and the tree, the following three factors must be considered in the following order:

Age. Big and old trees have a large root system, sometimes far beyond the dripline of the tree. To specify a certain distance from the tree to the pipe is difficult since different species and different soil conditions give different root systems. A rule of thumb given by Yingling et al (12) is not to trench closer to the tree than where the root is 2 inches thick. Hetherington (2) states that roots should never be cut if the diameter of the root is larger than 0.5 " where the trunk is under 6" and 1 inch where the trunk is over 6". A comprehensive survey on trenching near trees is given by Miller and Neely (5).

Tree species. The tree species is of great importance in determining how close to the tree the pipe can be constructed. Even if the tree has good soil conditions, a tree with an active and fast- growing root system can grow far beyond the original tree pit.

Soil volume. The volume of rootable soil may be of importance for the distance between the tree and the pipe. If the pipes are laid too close to the tree, drainage conditions and other soil parameters may be changed.

There is always a risk that a tree will fall in the future if lateral roots are cut. Tree stability is lowered and there is a risk of getting a storm-felled tree. The tree may survive for a period but soon it can become a hazard tree.

There are three important points to make before a tree is planted to avoid unnecessary expenses in the future due to root intrusion: 1) *Sewer lines*. Where are the different types of sewer systems installations situated? 2) *Type and age*. What material is the pipes made of? Do they have seals in the joints, and if they have, what are they made of? 3) *Condition*. The inside condition of the pipe plays a crucial role in deciding the extent of root intrusion. The best way to judge the condition is by TV-inspections inside of the pipe.

Where Root Intrusions have Occurred

When root intrusion has occurred with operational disturbance as a result, immediate measures must be taken. This is most important for sewer lines and combined sewers (both sanitary sewer water and storm water). A common problem is that the service connection from the sewer line to the house is plugged by roots and there will be problems with the drainage from one home. Another and more serious problem is when the main sewer line is plugged by roots which causes problems for many people at the same time. There is good reason to have a continuous inspection of the sewer system and plan for measures to be taken immediately when there is an indication of root intrusion.

Investigate before decisions are made. To remove existing trees and replant with new more suitable species might, in some cases, be a good idea if the design of the area is not dramatically altered. When measures have to be undertaken in areas where there are root penetrations, it is important not to generalize. Make an investigation where the whole problem is exposed and compare the technical, aesthetical, cultural and economical aspects.

A guideline (9) for every investigation can be as below:

1. Find out the costs and technical problems.
2. Identify the present measures.
3. Field investigation. A survey of soil conditions and an inventory of plant species.
4. Assessment of the plant value.
5. Suggestions of different modes of action: a comparison of economic, technical, ecological and aesthetic consequences.
6. Discuss the different solutions with all parties involved.
7. Decide.

To get as efficient action against the roots as possible, independent of method, the best results are achieved if a TV-inspection is made before taking action. This will reveal the extent of the intrusion and enable choice for the right method for solving the problem.

No-Dig Methods

Today there are a number of alternatives to the traditional digging and renewal of the pipes (11). When choosing one of these 'No-Dig Methods' it is important to know which method fits a specific situation best. The present condition of the pipe, the expected waterflow and the external conditions in the area have to be analyzed. These analyses should be done by an expert in the matter before the decision is made on which method is to be used.

Root cutting is the most common method to control roots in pipes. The root cutting is done mechanically with a knife or a steel wire around the inside walls of the pipe. Root cutting has been found to be only a temporary solution. In some areas with intense root intrusion the cutting has to be repeated every year. This will lead to high maintenance costs. Repeated trimming of roots on the same length of pipes leads to a greater risk of thinning out or damaging the pipe material. This can lead to a deterioration in the general condition of the pipes on the particular stretch.

High pressure flushing in pipes, which normally is done in combination with root cutting, may encourage root penetration to increase in intensity as it may wash away the seal in pipe joints. The risk for this is greatest when the seal is made with yarn and cement.

The most obvious reason for regrowth of roots in the pipe may be that when the roots are pruned they form new and even more fine roots. This method stimulates root growth. This is what happens when engineers develop technical methods without knowing anything about biology.

Pipe rehabilitation methods are presented below without any details.

- *Penetryn Method.* Penetryn is used to seal leakage with a liquid dual-component acrylic gel which is pumped into the joints. This method cannot be used where roots already have penetrated the joint.
- *Pipe Sliplining.* The method involves a 10-12 m long plastic pipe which is welded together and either drawn or pushed through the existing, defective pipe. This is a very good method that leaves no joints where roots can penetrate.
- *Short Pipe Sliplining.* Short pipe sliplining means that pipes are joined piece by piece with rubber rings and sockets. They are then drawn or pushed from the coffer dam through the old pipe. Since there are joints there is always a risk for root intrusion.
- *Pipe-cracking.* Pipe-cracking makes it possible to replace existing pipe-lines with new pipes of the same or larger dimensions. This is the best way to replace a pipe with a lot of problems.
- *Relining.* A liner of acid-resistant polyester fiber impregnated with resin is placed into the defective pipe with the help of water pressure. When the liner is in position, it is cured with hot water which gives a rigid and hard-wearing new pipe. This is a good method but there are indications that roots may grow between the old and the new pipe. When the root reaches the end of the weld, it grows into the new pipe.

Root Intrusion in Malmö, Sweden.

Root intrusion in sewer pipes has been very well studied in the city of Malmo (4, 8). The most intrusions have been onto the smaller dimension pipes, 225 - 400 mm (9). The larger dimension pipes are more often used as main sewer lines or trunk sewers. These pipes are usually deeper and the roots may have problems reaching them.

Data show that pipes laid down in the 1950s and earlier, with the exception of the 1930s, have the largest number of root intrusions (4). The explanation may be that during these years the sealing material was made of yarn and cement. During the war the quality of the material was poorer because of rationing. Why there were so few root intrusions in pipes from the nineteen-thirties is hard to explain. One reason may be that areas constructed during this period have not had so much vegetation near the pipes. During the 1950s the less effective plastic seals were introduced and since the 1960s rubber seals have been used. For further details see Rolf and Stål (6).

The large number of root intrusions in pipes from the sixties may be explained by the boom in the construction industry. Construction work was fast and cheap, which created negligence during the work. There is also a hypothesis that the number of root intrusions in these pipes will increase when the trees get larger.

Tree roots cause damage to sewer pipes. In the City of Malmo there is an annual expense of up to SEK 6 million ($750,000) / year because of the roots. The extent of root intrusion varies depending on the type of sewer and the type of vegetation and therefore all sites must be treated individually.

Three sites in the City of Malmo were inspected to find out alternative measures for each site. Every site was inspected to assess the technical, aesthetical, cultural and economical values. At one site with willows, the trees were replaced with less aggressive species and the expenses were shared between the wastewater works and the park department. At another site the aesthetical value was so high that the pipes were relined with a material that eliminates root intrusion problems in the future. At the third site, where trees were growing on private ground, it has still not been possible to reach a decision since the houses have to share the expenses of relining or taking the trees away.

During this project there was a discussion between the water works and the park department when every decision was taken. This is considered to be a prerequisite for good solutions to the problem.

Ecological Storm Water Handling

By concentrating on ecological storm water handling, in certain cases it should be possible to avoid the problems of root penetration in pipes. When ecological storm water handling is applied, one tries to imitate the natural cycle of the water. The term ecological handling of storm water comprises the concepts “local handling of storm water” and “open equalization of storm water/rain water (7)". When building new areas or when the system for storm water is being renovated or replaced, ecological storm water handling is an interesting alternative.

In short, “local handling of storm water” means that the water is infiltrated into a surface covered by vegetation or that the water is led into gravel pits whence it percolates into the surrounding soil. Percolation chambers, however, are not a good solution if one is searching for alternatives to decrease the problems of root penetration, as there are cases where roots have invaded pipes in these chambers (3).

The term open equalization means that the flow of rain water is evened out and delayed in open reservoirs, such as ditches and dams, often placed in parks or other green areas. Stahre & Larsson (7) give the following examples of constructions for open equalization: naturally soft surfaces, specially constructed infiltration beds, open ditches, open dams, prepared flooding areas with temporary dams, artificial marshlands/wetlands, root zone constructions, and alder swamps. These types of solutions take space, but have the positive effects of cleaner storm water, beautified surroundings,and lower investment and renewal costs. It is often possible to integrate constructions for open equalization even in built-up areas, although they must be adapted to local conditions.

There are many examples where open storm water systems have improved the landscape in parks and residential areas. The aesthetic value can in itself be a strong enough argument to choose streams, dams etc. instead of buried drainage systems. The design of the site is important however, both from an aesthetic point of view and bearing in mind the risk of children falling into the water. Children are tempted to play by water and this can help improve the play environment, but it is important to make flat slopes and to choose a water depth such that the risk for drowning is minimized. Where these demands have not been met, the water has often been fenced in to stop the play of youngest children.

Summary

Problems with tree roots and sewers are common in most cities. The problems that arise can normally be attributed to one of the following operations: trenching near existing trees, trees are planted close to existing sewer pipes, or when roots have to be taken away from sewer pipes. To be able to solve these problems, it is essential to find solutions that all parties involved will agree to and will do together. This means that the arborist must understand what the people from the water works are talking about. It is important to know their language and professional terms to be able to understand how a sewer system is constructed. There must be a give-and-take from all parties.

Willows and poplars are the trees that cause the most problems with root intrusion into sewers. In some cases these trees must be avoided and even removed because they cause major expenses to society and often can be replaced by other, less aggressive species.

Methods to solve problems of root penetration in sewer pipes caused by trees are often approached with traditional thinking and technical know-how. In the short run, this often gives good results, but pays little consideration to the biological conditions on the site, which is why it may aggravate the problems. It is therefore necessary to improve the knowledge of the people involved, in the biological effects of different courses of action, and also inform them about alternative solutions to the problem of root penetration in pipes. By finding solutions that take into consideration the economic, technical, aesthetic and ecological aspects of the problem, those in charge may choose a more appropriate course of action.

Ecological storm water management has primarily been used for the removal of pollutants and is often less expensive than conventional piping systems. The use of wetlands and ponds can add aesthetic value to urban areas as well as remove storm water. If it is possible to channel it via open ditches and ponds rather than using pipe systems, then the root penetration problem is automatically solved.

Acknowledgments

This work was supported by the Swedish Council for Building Research (BFR) and the Swedish Water and Wastewater Works Association (VAV).

Literature Cited

1. Cutler, D.F. and I.B.K. Richardson. 1989. Tree Roots and Buildings. Longman Scientific & Technical, Harlow.
2. Hetherington, R. 1994. Trenching and street trees: Local co-operation and liaison. Arboric. J. 18: 237-245.
3. Jansson, E., B. Lind and B. Malbert. 1992. Lokal dagvattenhantering. Erfarenheter från några anläggningar i drift. VA-FORSK. Rapport 1992-09. (in Swedish)
4. Lidström, V. 1994. Change in structural condition in sewer pipes. Proceedings NODIG 94, Copenhagen, June 1994.
5. Miller, F.D. and D. Neely. 1993. The effect of trenching on growth and plant health of selected species of shade trees. J. Arboric. 19: 226-229.
6. Rolf, K. and Ö. Stål. 1994. Tree roots in sewer systems in Malmö, Sweden. J. Arboric. 20(6), 329-335.
7. Stahre, P., and T. Larsson. 1993. Dagvattnets roll i framtidens stadsplanering. Stadsbyggnad, nr 4. (in Swedish)
8. Stahre, P., A-C Sundahl and Lidström V. 1994. Valedningars kondition. VA-FORSK. Rapport 1994-01. (in Swedish)
9. Stål, Ö. 1992. Trädrötter och ledningar (Roots and pipes). Svenska vattenoch avloppsverksföreningen, VAV. Report 1992-14. (in Swedish)
10. Stål, Ö. 1995. Trädrötter och avloppsledningar. Enkätundersökning av rotinträngningars omfattning i avloppsledningar hos sveriges vaförvaltningar. Svenska vattenoch avloppsverksföreningen, VAV. Rapport 1995. (to be published in Swedish)
11. VAV, 1989. VAV Publikation P66. Svenska vattenoch avloppsverksföreningen. Stockholm. (in Swedish)
12. Yingling, E.L., C.A. Keeley, S. Little and J. Burtis Jr. 1979. Reducing damage to shade and woodland trees from construction activities. J. Arboric. 5: 97-105.

Interactions Between Tree Roots and Buildings

David F. Cutler

Data on tree roots derived from survey work related to a root identification service and from exposed root plates following severe storms in south-east England in 1987 and 1990 are discussed in the context of tree roots and buildings. The generally shallow depth of structural roots of many trees, and the relatively narrow diameter of roots at 1-2 meters from tree trunks are considered in relation to ground work near trees. The possible misinterpretation of tree root spread data in relation to insurance risk is emphasised.

Trees in urban settings are frequently blamed for damage to buildings or structures in which cracks are found, even when many other factors could be the sole cause of, or could have a considerable part to play in, the damage recorded. Trees have also been the focus of complaints when drains have become blocked by roots, despite the fact that drains probably have to be leaking before roots can enter them. In addition, those responsible for underground services have been concerned about the proximity of tree roots, which they thought may somehow damage the utilities. This means that trees in towns and cities have to survive a hostile environment, not only in their growing conditions, but also in the attitudes of house owners, insurers and those providing the utilities.

This paper summarises knowledge about the morphology of tree root systems and is mainly based on research done at the Jodrell Laboratory, Royal Botanic Gardens, Kew over the past 25 years. Published information on root spread and their risk to buildings often is misinterpreted, sometimes through vested interest but often through treating data uncritically. This is despite carefully worded warnings in these articles on what is known about tree roots and the relationships between trees and buildings. Some of the main misconceptions about tree roots are discussed, and possible lines of research are indicated that could clarify matters further.

Tree Root Research at Kew

The Anatomy Section of the Jodrell Laboratory runs an identification service for material of plant origin ranging from antique furniture and archaeological remains through food contaminants to forensic work. Tree root identifications figured in this work, and have represented a regular

Dr. David F. Cutler, Head of the Plant Anatomy Section, Jodrell Laboratory, Royal Botanic Gardens, Kew, Richmond, Surrey, TW9 3DS, England.

component of the enquiries. In the early 1970's when in the UK insurers of buildings began to include cover for subsidence, the number of roots sent for identification increased dramatically. Previous to this, in general, owners of property that subsided had to pay for remedial treatment themselves, unless others could be found to bear the blame.

It can be important to establish if a tree might be involved, by excavating for roots near the point of subsidence. If roots are present, the tree to which they belong has to be found, since the tree owner is responsible for any damage caused by its roots. The invasion of land belonging to someone else by tree roots is described as 'root trespass.' Insurers of buildings which had subsided and cracked would prefer to find roots of a tree that is not on the property they cover, so that they can offset their loss.

Trees have been blamed for cracking that may have arisen from a variety of causes, since they are an easy and readily seen target. Many street trees came under scrutiny, and the insurers of municipalities became increasingly involved in meeting claims. The main winners in the situation were rarely the trees. This is, unfortunately, still true.

It was clear that since large numbers of roots were being identified, there was the potential to conduct research into root spread of a variety of trees. A survey was started. Survey cards were designed, and sent with the identifications to customers who had initiated the enquiries. These were normally loss adjusters acting for insurers. Among other things, the distance of the tree from the excavation at the property and the tree height were asked for, as was the soil type. Previous to this, most data on root spread had been obtained by excavating root systems of a rather small number of species (e.g.10). Excavation is a costly and time consuming activity.

The root spread survey arose, then, from the identification service, and analysis of 2600 cards led to publication of 'Tree Roots and Buildings' (6). It is interesting to note that the British Insurance Association, as an alliance of insurance companies was then called, provided money to speed up the analysis, leading to the publication of this book. A second, revised edition of the book was published (7). This incorporated data on a further 11,000 trees and 1,300 shrubs resulting from identifications of roots at a consultancy run by I B K Richardson.

The addition of the new data did not change the conclusions reached about root spread in the first edition. The accuracy of the interpretation of the original data was reinforced. This is important, since the degree and extent of cover of survey work of this type depends on chance, unlike directed survey research, where sampling can be made according to accepted statistical standards.

A further product of the identification service was a volume aimed at making available information amassed on tree and shrub root anatomy, as an aid to root identification. This Root Identification Manual of Trees and Shrubs (8) covers 250 species or varieties in 129 genera. They are all capable of growing out of doors in the British Isles.

It had been recognised for a long time that virtually all root action leading to subsidence damage to buildings was secondary in nature. That is, direct physical pressure exerted by roots was very rarely the cause of cracking. A tree would have to be very close indeed to the structure for such damage to occur. Pavements (sidewalks) and boundary walls with little or no foundation can be lifted by large shallow roots of, for example, *Populus* species. Buildings with adequate foundations, standing some distance from trees, did not normally suffer from such direct action.

It was thought that the problem was largely confined to soil types that shrank considerably on dehydration (3,4,9). These are mainly certain types of clay and some peaty, fen soils in the UK. Under conditions of water deficiency, when tree roots remove water from shrinkable soils, they accelerate the drying and shrinking process.

In October 1987, a very severe storm struck the south-east of England. It followed a wet period, and deciduous trees were still in full leaf. Over 15 million trees were estimated to have blown down. Over 800 fell at Kew alone, and in the satellite garden at Wakehurst Place, several thousand trees were either blown down, or large parts of the crowns were damaged. Many trees fell exposing their root plates. These plates were generally found to consist of the main roots radiating from the trunk, extending to the region where rapid taper ceased, and where the laterals had reached the dimensions that they kept for several to many meters. Root plates ranged in radius from one to 2-3 or more meters. Roots of trees at Kew and Wakehurst Place were measured and photographed, and other details of the trees were recorded, including height and diameter at breast height.

We were invited by the Department of the Environment, who set up 'Task Force Trees' in 1987 to help deal with the catastrophe, to embark on a survey of wind-blown trees for the area of England that had been affected. Survey questionnaires were designed by the team to help in collecting the required information. Form A was designed for information on individual trees in such a way that members of the general public could play an active part in data collection. The second, (B), was designed for professionals, and covered groups of trees of the same species from woodlands and forest plots. A letter in 'The Times' inviting public participation produced a remarkable response, and in total 886 trees were recorded on form A and 3625 on form B. Since form B also contained details of the proportion of trees that did not blow down in the woodland stands reported on, the total number of trees examined for this set of forms was about 8000.

Details of the survey were published in a number of papers, and the main analysis was reported as *The Wind Blown Tree Survey: Analysis of Results* (5). Forms A and B are illustrated in that paper. The results of the work on tree root plate morphology were also published (11). Examples of the data obtained and the main conclusions derived from them are discussed below.

In January, 1990 a further storm struck, blowing down many more trees, some of which had been left exposed by the previous storm. Additional data were recorded.

Data in 'Tree Roots and Buildings'

The second edition of this volume contains information on the root spread of 18 genera of trees commonly reported to be involved in damage to buildings. Lists are given of another 28 genera of trees and 45 or so shrubs that have occasionally been of concern or for which roots have been identified close to failed foundations. Although the maximum root spread encountered was reported for each species (or genus, when it was not possible to be more precise in identification of the roots), this is not thought to be the most useful information. Of far more value for those concerned with a balanced view of trees and buildings, is the distance within which 50% of the enquiries arose. If this figure poses problems for the insurers, then they are at liberty to examine the tables and graphs and determine the point at which 'acceptable' risk might be taken.

Unfortunately a rational view from the insurer's standpoint does not seem to accord with that of the trained arboriculturist. This is bad in itself, but becomes much more untenable when the soil type is ignored. It is quite clear from the book (see also Table 1, this paper) that 96-100% of the instances when the 18 most recorded genera of trees were thought to have been contributors to damage to structures they were growing on shrinkable clay soils. The results have often been extrapolated to all clay soils, in a scientifically unsound way.

The normal soil drying profile of tree root systems, where it is known (e.g. 1), is also largely ignored by those who have become over-anxious about the potential effect of trees on buildings. Table 1 shows for selected genera, the percentage of cases reported for shrinkable clay soils in relation to all cases reported, and the percentage of all cases where roots were reported to have entered drains. Table 1 also shows the distance within which 50% of damage has been recorded for 18 genera. It also includes the total number of cases for the research period in which each genus was thought to have been implicated. Table 3 gives the distances within which 75% and 90% of damage was recorded, and cites the maximum root spread noted for 17 genera. (The data for *Chamaecyparis* have been included with those for *Cupressus* in this table.)

Unfortunately, insurers are inclined to take the extreme figures, to minimise their risk. This is why it is so important that a qualified arboriculturist should give guidelines on planting distances in new planting schemes. Not only can the characteristics of particular species be taken into account, but proper consideration of the soil type can be given. There is a move by insurers towards consulting the Geological Survey about soil type in the UK. This is better than nothing, but if the post code (zip code) is the smallest unit

Table 1. Data about tree genera derived from enquiries (7).

Genus	No. cases	% on shrinkable soil	% in drains	Distance from buildings % damage recorded (m) 50%	75%	90%	Max root spread (m)
Acer	1098	99	9.6	6.0	9.0	12.0	20
Aesculus	386	98	11.0	7.5	10.0	15.0	23
Betula	217	100	5.5	4.0	7.0	8.0	10
Crataegus	615	99	1.6	5.0	7.0	8.7	11.5
Cupressus	407	100	2.5	2.5	3.5	5.0	20
Fagus	154	100	>1.0	6.0	9.0	11.0	15
Fraxinus	1007	99	4.5	6.0	10.0	13.0	21
Malus, Pyrus	80	96	2.0	4.0	6.0	8.0	10
Platanus	1467	100	7.5	5.5	7.5	10.0	15
Populus	1165	99	24.0	11.0	15.0	20.0	30
Prunus	860	100	4.5	3.0	6.0	7.5	11
Quercus	1546	100	3.5	9.5	13.0	18.0	30
Robinia	411	100	>1.0	7.0	8.5	10.5	12.4
Salix	787	100	18.5	7.0	11.0	18.0	40
Sorbus	303	96	>1.0	5.0	7.0	8.5	11
Tilia	1112	100	1.0	6.0	8.0	11.0	20
Ulmus	316	100	>1.0	8.0	12.0	19.0	25

area worked on, this could lead to unfairness on the boundaries of the shrinkable clays.

It is clear from reading the book and consulting Table 1 that members of some tree genera rather than others potentially pose much more of a risk on shrinkable clay soils. These tend to be forest or park-land trees, which can eventually grow to be completely out of scale with the buildings around them. For example, plane (*Platanus*), poplar (*Populus*) and oak (*Quercus*) may become very large trees. Others, like birch (*Betula*) give a much reduced cause for concern.

The common view that roots extend as far as the drip line of the branches is patently untrue. (Even a non expert can see this for the Lombardy poplar!) Table 2, using data from (6), shows for selected genera the relationship between the percentage of cases of damage relative to all trees reported to have caused damage and the estimated percentage of planting of the genus relative to all street and garden planting. The significance of the figures is as follows. If the reported damage figures are close to the planting frequency figures, then nothing exceptional is happening in relation to that genus. If the damage figures are much lower than the planting frequency figures (e.g. *Betula, Sorbus*), this indicates that a genus is relatively safe to plant, within the distances suggested by the other tables, whereas if the damage frequency figure is much higher than the planting frequency, the genus could

Table 2. Percentage of cases of damage by members of particular tree genera in relation to estimated relative frequency of street and garden planting. Data derived from (7).

Genus	% of Damage	Est. % of total planted
Acer	8.3	13.9
Aesculus	2.9	2.2
Betula	1.5	4.9
Crataegus	4.6	3.5
Cupressus	3.0	10.0
Fagus	1.0	1.0
Fraxinus	7.5	3.0
Malus, Pyrus	5.7	7.1
Platanus	11.0	13.7
Populus	8.7	3.0
Prunus	6.4	14.7
Quercus	11.5	2.1
Robinia	3.0	1.8
Salix	5.7	4.5
Sorbus	2.3	9.4
Tilia	8.2	6.4
Ulmus	2.0	NA

pose particular problems (e.g. *Quercus*).

We have no direct evidence from the survey, but from discussions with professionals investigating subsidence throughout the period of the survey, it appears that drains were probably leaking before roots entered them. Roots often grow near to intact drains and other service pipes when these are cooler than the surrounding soil. Water condensing on them (as it does under paving slabs) would be attractive to roots. Roots grow towards an increasing water gradient, and appear to need at least a continuous surface film to direct them to the water. Detection by roots of a water supply inside an intact pipe would require sensing methods unknown to plant physiologists.

Cutting trees back to reduce water demand appears to be an effective if undesirable practice. Not all species will respond satisfactorily to cutting. Some information is given by Cutler and Richardson (7). Complete removal of trees on shrinkable clay soils can lead to soil recovery and heave. The relative costs of tree crown volume management and repairs to a structure following heave have to be balanced before a decision can be taken. Because foundations differ in depth, and buildings have different types of construction, there is no simple advice to be given in instances of tree related subsidence. A professional team is needed, in which the arboriculturist plays a key role, for sound decision making.

Data from Research Following Storms

This present paper concentrates on information about root structure from Gasson and Cutler (11) and Cutler, Gasson and Farmer (5). These papers contain other information on, for example, soil types related to wind blow frequency, which are outside the brief of this present work.

Table 3 shows how many (and what percentage of the sample of) trees had roots within particular size categories at 1 m from the trunk. Of particular interest is the fact that 27% of the sample consists of trees with roots of 5 cm diameter or less, and 66% of trees had roots of no more than 10 cm diameter at this distance from the trunk. This is very important, because contractors working in the ground near to trees would in all probability dismiss roots of these diameters as having little significance. They are, of course, the main structural roots.

Of similar importance is the information on root depth, which was measured at the root plate. The structural roots of many of the trees that blew over were demonstrated to be much shallower than conventional wisdom would suggest. There is no reason to think that this was a biased sample. Table 4 defines four depth categories and, for each, presents combined numbers of trees recorded on the survey forms A and B, together with the percentages of the total sample that these numbers represent. The following points are of key interest: 96.5% of trees had structural roots exposed at the root plate that were no more than 2 m deep, and 44.4% had roots 1 m or less

Table 3. Root diameter data from individual trees after the 1987 storm. Data derived from (5).

Max. root diameter at 1 m from trunk (cm)	No. of trees	%
Fibrous	27	4
2-5	177	23
5-10	301	40
10+	255	33

Table 4. Root depth categories for all trees measured after the 1987 storm. Data derived from (5).

Root depth	Combined A & B data	%
<0.5 m	528	15.2
0.5 - 1.0 m	1083	31.2
1 - 2 m	737	50.1
>2 m	121	3.5
Total	3469	100.0

deep. This means that even shallow trenching close to trees could sever a significant proportion of structural roots.

Root plate diameter tended to be related to the distance from the trunk at which rapid root taper ceased. This appeared to be the nearest 'weak point' to the trunk where roots snapped. The diameter reached at this region would continue more or less constant, in large trees, for many meters. There is, of course variation between genera in this respect. Table 5 summarises this information. Notice that conifers are no different in their size range from broadleaf trees, with the exception that *Picea* just beats *Fagus*. This information is of value to those wishing to work near large trees, since it indicates where damage might be caused to root plates, as opposed to the extension system of roots. There is, of course, considerable variation in the root plate radius within the genera.

Table 6 shows the root plate radius categories, and numbers and percentage of trees in each. It is instructive to show separately and in combination the information obtained from the two forms, A and B. That

Table 5. Maximum root plate radius recorded following the 1987 storm for selected genera (Data derived from 5).

Genus	Max. root plate diameter	Genus	Max. root plate diameter
Picea	8.4	*Pseudotsuga*	4.0
Fagus	8.0	*Larix*	3.7
Fraxinus	6.0	*Cedrus*	3.5
Castanea	6.0	*Carpinus*	3.3
Abies	6.0	*Platanus*	3.2
Quercus	5.0	*Malus*	3.0
Betula	5.0	*Prunus*	3.0
Acer	5.0	*Salix*	2.6
Cupressus	5.0	*Chamaecyparis*	2.4
Pinus	4.1	*Aesculus*	2.3
Populus	4.1	*Sorbus*	2.1
Tilia	4.0		

Table 6. Root plate radius categories (data derived from 5)

Root	(A) Trees		(B) Trees		(A) & (B)
radius	No.	%	No.	%	%
<0.5 m	8	1.0	62	2.2	2.0
0.5 - 1 m	113	14.8	1608	58.0	48.6
1 - 2 m	321	41.9	935	33.7	35.5
>2 m	324	42.3	168	6.1	13.9
Total	766	100.0	2773	100.0	100.0

Table 7. Root type frequency, recorded by Cutler and Gasson (5) following the 1987 storm. LD = lateral and droppers, L = lateral, S = sloping, TL = tap and lateral, V = vertical and TT = two-tiered.

Frequency	LD	L	S	TL	V	TT	Total
Forms (A) & (B)							
Total	2240	1169	520	100	97	80	4206
%	53.3	27.7	12.4	2.4	2.3	1.9	100

derived from form A is for solitary trees, and that from B for trees in plantations or woodlands. Root plates are, on average, narrower in plantation trees than in those grown apart from others. For example, 42.4% of solitary trees have root plates wider than 2 m in radius but in plantations the figure is only 6.1%.

The storm also provide interesting information on root types. Of most significance is the observation that only a small proportion of trees has tap roots (2.4%). The vast majority (81%) have either lateral or lateral with droppers (sinkers). The results are set out in Table 7.

Conclusion

The research on tree roots in relation to buildings conducted at Kew has been summarised above, and salient features discussed. It is clear that there are still many gaps in our knowledge. For example, little is known about the dimensions and depth of structural roots of large trees from between about 4-5 m from the trunk to their growing extremities. It can be inferred that for many species, they probably remain at about the same depth as where they were recorded near the root plate.

Of more pressing importance is the need to know how much of a root system can be severed before death or instability will ensue. This is particularly difficult to find out, since apart from a few specimen trees in level ground on soils with uniform consistency, roots rarely radiate evenly and at equal spacing from the trunk. Designing any cutting experiments without knowing where the roots are is fraught with difficulty. Yet if we are to give sound advice on the probable effect of trenching, say for cable TV, at various distances from trees, we can at present give only an educated estimate. Long term observation needs to be made of trees that have been subjected to digging or trenching near to them. The contractor may have moved on (or become bankrupt) before the effect of damage caused to trees becomes evident. It took three to four years, for example, for one group of hybrid poplar trees that remained standing after the storm to eventually die, but whose roots were snapped in the ground. The work of Watson (this volume) on experimental trenching near trees should add significantly to our knowledge in this area.

Another controversial and as yet inadequately researched area is the provision of permanent or temporary unpaved car parking under mature trees. It is to be anticipated that the trees would suffer, but as far as I am aware, there is no published information available. This could be easily remedied in situations where 'control' areas already exist. Guidelines for those constructing near trees provide for an area around a tree that should be fenced off from traffic movement or the dumping of materials (2), so clearly there is a perceived problem.

There is, of course, active research on root barrier material, and the effects of roots on soil drying are being measured. In each of these areas of work still much remains to be done. There will be no shortage of research projects for those interested in trees and buildings, or in the advisory work for arboriculturists in the foreseeable future. What we need to do now is consolidate the current knowledge base, and follow up some of the pressing needs for work, so that the professional advice that is given by tree specialists remains of the highest calibre.

Literature Cited

1. Biddle, G. 1983. Patterns of soil drying and moisture deficit in the vicinity of trees on clay soils. Geotechnique, 33 (2): 107-126.
2. British Standards Institution. 1991. Guide for trees in relation to construction. BS 5837.
3. Building Research Establishment. 1965, 1972, 1976. Soils and Foundations, 1-3. Digest 63, 64 and 67. BRE.
4. Building Research Establishment. 1980. Low-rise buildings on shrinkable clay soils, 1-3. Digest 240, 241 and 242. BRE.
5. Cutler, D.F., Gasson, P.E. and Farmer, M.C. 1990. The wind blown tree survey: Analysis of results. Arboric. J.14(3): 265-286.
6. Cutler, D.F. and Richardson, I.B.K. 1981. (1st ed.) Tree Roots and Buildings, Construction Press, Harlow, Essex UK. 94 pp.
7. Cutler, D.F. and Richardson, I.B.K. 1989. (2nd ed.) Tree Roots and Buildings, Longman Scientific and Technical, Harlow, Essex, UK. 71 pp.
8. Cutler, D.F., Rudall, P.J., Gasson, P.E. and Gale, R.M.O. 1987. Root Identification Manual of Trees and Shrubs. Chapman and Hall, London, UK. 245 pp.
9. Driscoll, R.C.M. 1983. The influence of vegetation on swelling and shrinking of clay in Britain. Geotechnique 33 (2): 93-105.
10. Fayle, D.C.F. 1968. Radial growth in tree roots: distribution, timing, anatomy. Tech. Rep. No. 9. Faculty of Forestry, University of Toronto.
11. Gasson, P.E. and Cutler, D.F. 1990. Tree root plate morphology. Arboric. J.14(3): 193-264.

Climbing Plants on Walls – Advantages and Disadvantages

Palle Kristoffersen

For many years climbing plants on walls have been considered to have a negative effect on buildings. The most common assumption was that the plants cause walls to become damp. Another assumption was that the plants could damage the brickwork by growing roots in the bricks and in the mortar joints. Another major problem was considered to be Hedera's negative phototropism and its ability to grow into existing openings or cracks, and consequently the diameter growth might give problems. A research project based on a literature study and an inspection of more than twenty-five Danish examples showed that problems occurred only in cases with no or insufficient maintenance.

In Denmark and other European countries, different kinds of climbing plants on walls have been considered to have a negative influence on buildings. When this project was initiated in 1992, there was a general resistance among the building trades and the insurance companies to placing climbing plants on walls.

These assumptions were supported by several Danish official instructions to building maintenance officials that warned against climbling plant on walls because they damage buildings. However, this study has proved that it is difficult to find cases where climbing plants on walls have actually damaged the brickwork. The negative assumptions were presumably based on wrong and ill–founded information.

The negative assumptions of building technicians are in sharp contrast with the opinion of landscape architects and landscape gardeners who in general believe that climbing plants on walls do not damage brickwork if the planning, establishment, and maintenance are carried out correctly and in accordance with the construction and management of the building.

The differences of opinion have had the consequence that climbing plants on walls are now being established along with removal of old, large climbing plants (Fig. 1). The removal of old plants was often enduced by the insurance companies who threatened to renounce building insurance.

This paper includes a description of climbing plants and their characteristics, and a description of brickwork and its characteristics. The descriptions are based on understanding of the relations between brickwork and plant growth. This has been necessary because the assumptions of influence of plant growth on buildings (especially the negative) are based on insufficient

Palle Kristoffersen is Research Landscape Architect at The Danish Forest and Landscape Research Institute, Department of Park and Landscape, Hoersholm Kongevej 11, 2970 Hoersholm, Denmark.

Figure 1. Climbing plants can be a part of the architectural expression. If they are established and maintained correctly a Danish study proved that there are many advantages and only few disadvantages as regards climbing plants. The picture shows *Parthenocissus tricuspidata*, Japanese Ivy.

knowledge of the anatomical and physiological characteristics of the plants as well as the differences between individual plant species and the physical characteristics of brick walls.

Climbing Plants and their Characteristics

Climbers are lianas and are characterised by their long and flexible stems. They are unable to support themselves and therefore have adapted themselves to utilize different kinds of supports in order to reach the light (13). Climbing plants can be self fastening, i.e., they can fasten directly on the wall with special fastening organs. The most frequent and largest group of climbing plants on walls are self-fastening climbers, especially *Parthenocissus tricuspidata* (Japanese ivy) and *Hedera helix* or *H. hibernica* (ivy).

In nature, the plants fasten on rocks and stems of other plants. They have developed clasper roots or clasper discs. Examples of plants with clasper roots are *Hedera* spp. (ivy species) and *Hydrangea anomala* ssp. *petiolaris* (climbing hydrangea). The clasper roots are root–like fastening organs developed on the vegetative shoots on the side opposite to the light, i. e., against the support (9). Clasper roots are developed in the same way as the ordinary nutrition roots. However, the clasper roots lignify fast and become physiologically inactive, i.e., they do not transport water or nutients (3).

The clasper roots are established on the vegetative shoots as a consequence of low light and an adequate amount of water (6). Therefore, among other reasons, it is difficult for *Hedera* to establish clasper roots on walls exposed to sun. In contrast to nutrition roots, clasper roots are undivided,

Figure 2. The *Parthenocissus tricuspidata*– Japanese Ivy climbs with clasper disc. The climbing threads establish small (1–2 mm in diameter) clasper disc which occur 4–8 at a time. Upon fastening the disc thread tightens into a spiral and pulls the plant close to the wall. The fastening is established when the disc cells produce a viscid secretion that hardens on the support with exposure to air. After the fastening the climbing treads get physiological inactive.

8 –10 mm long, and placed in 2 cm long double–rows on the lower side of the shoot.

Studies from 1895 and 1910 show that the fastening function is established when fine root hairs from each clasper root grow into the pits and irregularities of the support. The strength is determined by the number of root hairs with terminal thickenings. The thickening prevents the root hair from slipping out of the support. The root hairs are small extensions of the outer cell layer on the rhizoderm (3).

If *Hedera* clasper roots are exposed to constant moisture and darkness during their early growth (< 0.5 mm), the development process may change and the clasper roots could revert into nutritional roots (3). Consequently, with penetration to permanent moisture sources, *Hedera* acts as a pseudoepiphyte and is able to absorb moisture from a damp wall. It his not known what quantity of moisture is required for *Hedera* clasper roots to develop into nutritional roots.

Clasper disc climbers belong to the genus *Parthenocissus* (Japanese ivy and Virginia creeper)(Fig. 2). These climbing plants have tendrils with the ability to establish small, round clasper discs when they touch the support. The diameter of a disc is 1–2 mm and they occur 4 to 8 at a time. When fastening, the clasper disc threads tighten into a spiral, thereby pulling the plant very close to the wall. The fastening is established when the disc cells emit a viscid secretion that hardens to the support with air contact (3).

After the primary fastening the clasper discs grow further, and the new tissue adjusts to cracks and irregularities which strengthens the fastening effect. The clasper discs lignify at the end of the growth season, at the latest, whereby they become physiologically inactive (4).

A number of sources have stated that clasper roots and discs emit chemical substances that decompose the walls (7,8). In truth, an insignificant amount of yellow secretion is liberated during establishment and also an equivalent insignificant amount of acidic cell sap is found when the *Hedera* clasper roots die. The *Parthenocissus* clasper discs emit a secretion during the fastening process which contains acid but in very small amounts. Claims that clasper organs decompose brick walls have been challenged (1).

Twining plants are climbers that fasten as the shoots grow upwards encircling with a spiral. The plants seek support through constant circling of the top shoot. This is possible due to different growth rates of the various shoot sides. The leaves do not unfold until the shoot has found support. If there is no support the stems twine around themselves, extend several meters and climb on into trees or installations.

The rest of the climbing plants that require support have developed other related climbing functions. In general, climbing plants fasten through tendrils, sensitive petioles or other adapted leaf parts. Tendrils are adapted stems (13). On contact with a suitable support, the tendrils grip with a spiral growth and lignify. An example of tendril climbers is *Vitis vinifera* (ordinary grape). Leaf climbers fasten through twining petioles around the support (13).

Some climbing plants on walls have few or no specific climbing organs and need to be tied to a support. They are not included in this paper.

Growth Form and Site Demands

Climbing plants on walls can often be used at places where other plants cannot grow. These plants grow in a protected site, i.e., along a wall where the microclimate is perfect for the plant and the plant roots can develop along the foundation and, when the conditions allow, under the surrounding pavement which makes the soil volume much larger than it seems. In contrast to trees, climbing plants do not develop buttress roots since they hold on to other structures for support. Therefore, it has been observed that even large climbing plants do not destroy the pavement in the same way that trees do.

The fact that the plants grow along the foundation, and as for *Hedera* species, are evergreen, means that a considerable amount of water is absorbed, which contributes to keeping foundations and basements dry (11).

Climbing plants on walls have a very characteristic growth form. *Hedera* grows upwards and outwards in an angle of 45 degrees from vertical, while *Parthenocissus tricuspidata* grows upwards and outwards in an angle of 90 degrees (4). Sometimes *Parthenocissus* grows downwards in an angle of 45 degrees from horizontal. These characteristics are important for plant choice in relation to the individual design requirements for the plant density. For *Hedera* a much higher plant density is needed than for *Parthenocissus*, where one plant can cover several hundred square meters of brick wall and even grow around corners. However, the ideal condition for *Parthenocissus* is to plant one plant per brick section, split by downpipes. Our study shows that if the plants are kept on each side of the downpipes, the main stems will not deform the pipes.

The vegetative shoots of the self-fastening climbers seek darkness and grow away from direct sunlight and towards the support to which they fasten (4). This means that, in theory, the plant shoots might grow into cracks and openings. Therefore, the shoot diameter growth might cause problems to buildings. If there is no room for the diameter growth, there is a risk that plants will lift, for instance, roofing tiles.

The dense leaf cover of *Hedera* , and in summer of *Parthenocissus*, is the main reason that some people say that the plants have positive influence on buildings and some say that they have negative influence.

Parthenocissus establishes short shoots with leaves and flowers on the main stems. Over the years, the leaf cover grows more and more dense. It is rare when *Parthenocissus* plants loosen from a wall. *Hedera* grows in a different way (Fig. 3). When the several meter long annual shoots no longer have support, e.g., when the plant has reached the top of the building and when the shoots do not fasten as usual, the shoots will droop. The study shows that the drooping shoots encourage establishment of flowering shoots and a thick carpet is established. The thick, heavy carpet is the reason why *Hedera* plants sometimes loosen and fall away from the support. This can be avoided through regular cutting of loose shoots.

Relationship to Brickwork

In Denmark, many of the largest and oldest climbing plants on walls are found on traditional brick buildings constructed during the first half of this century. Very often climbing plants were planned as a part of the architecture. The opponents to climbing plants have claimed that these are the buildings that can be damaged.

Brickwork consists of bricks joined by mortar. The bricks account for approx. 75% of the frontal area and the joints of approx. 25%. In older buildings prior to the 1930s, the outer walls are mostly solid with thickness

Figure 3. If Hedera's several meter long annual shoots do not fasten on the wall they will droop. The drooping shoots encourage establishment of flowering shoots and a thick carpet of leaves and flowers can be established. If the carpet is not pruned regularly the weight from the carbet will make the plant loosen and roll off the wall.

of 1.5 times the brick length (350 mm) or more. The thickness is dependent on the height of the wall and the load at a given height. The joints contain mortar with hydraulic lime or cement added to increase strength. In new buildings the outer walls are mostly hollow with 0.5 times the brick length with mineral wool insulation in the cavity. Bricklaying has been done with a cement and sand mortar. These mortars are stronger than those used in older brickwork.

When exposed to water, brickwork absorbs water, since bricks and mortar are of a porous material with a huge volume of pores. By far most of brick pores are 0.15–1.0 μ (0.00015–0.001 mm) in diameter. However, the water evaporates quickly from the brick pore system. Tile bricks absorb little moisture from the air as they have practically no submicroscopic pores, e.g., with a diameter less than 0.0001 μ. Under normal conditions, the brickwork moisture content, including mortar, is approximately 1% by weight (1.75% by volume). Experience has shown that in periods with heavy rainfall the moisture content can be considerably higher, up to 10% by weight.

Plant cover on walls is not considered to influence wall evaporation, even though the plants provide shade and shelter. Further, the brickwork will not absorb moisture even though the relative humidity between the plants is very high.

At constant humidity and poor light exposure *Hedera* might develop secondary nutrition roots from the clasper roots on young shoots. This is found if deep cracks in brickwork or plaster occur. However, the literature has no examples of self fastening climbers penetrating walls.

The Danish Study

Our study deals with both deciduous and evergreen plant species. The chosen species are *Hedera helix* (ordinary ivy), *Hedera hibernica* (large–leaved ivy) and *Parthenocissus tricuspidata* (Japanese ivy). The study includes 22 localities. The localities have been selected according to the following criteria: 1) the buildings must be from the first half of this century or last century, 2) it is assumed that climbing plants have damaged the brickwork, 3) there is variation in the age and maintenance condition, and 4) preferably, the plants are facing north west and west. Furthermore, in 6 of the localities bricks and mortar samples were removed from the outer walls and tested for moisture content. The samples were weighed before and after drying to a constant weight at 110°C. Based on the weight loss, the moisture content was calculated (Table 1). After drying, the bricks were placed in water for 48 hours whereupon water uptake and density were measured.

The water content of the bricks was very low, all below 1% by weight. Brickwork exposed to a heavy shower can have a significantly higher

Table 1. Moisture content of the tests and brick water uptake at 48 hours water storage and density.

	Storey	Facing	Plant cover	Moisture content		Brick water uptake at 48 hours water storage		Brick density kg/m³
				Brick	Mortar	%-weight	%-Volume	
Brickwork with Ivy plant cover								
Ole Suhrsgade Copenhagen	3rd floor	North east	With	0.24	0.21	28	41	1482
	3rd floor	North east	Without	0.56	0.23	23	37	1576
Gasværksvej 2 Copenhagen	2nd floor	South west	With	0.40	0.24	21	34	1569
	2nd floor	South west	Without	0.45	0.31	22	34	1581
Århus University Building 161	3rd floor	South	With	0.05	0.11	16	28	1714
	3rd floor	South	Without	0.96	0.70	19	31	1628
Mejlgade 78b Århus	2nd floor	South	With	0.28	0.45	24	37	1518
	3rd floor	South	Without	0.13	0.30	23	35	1541
Brickwork with Japanese Ivy plant cover								
Snorresgade 5-7 Copenhagen	2nd floor	West	With	0.19	0.22	12	22	1849
	2nd floor	West	Without	0.57	0.23	14	26	1835
Århus University Building 110	4th floor	North	With	0.18	0.65	20	33	1642
	4th floor	North	Without	0.07	0.47	18	30	1662

moisture content at times. Front brick water uptake is approx. 20 - 35% by volume and the density often 1700 to 2000 kg/m^3. Thus some of the tested bricks have a moisture uptake above normal and a density below normal.

None of the results in this study indicates that *Hedera* causes an increased moisture content in the brickwork. To the contrary, measurements of moisture content in samples of bricks and mortar were very low; all below 1% by weight. In three cases, the moisture content was lower in samples taken under the climbing plants than those taken beside the climbing plants.

The organic material in the plant was, at the time of the survey, dry and loose and not further decomposed which indicates that it is permanently dry. Out of all 22 building walls covered with plants, not a single case was found with moisture damage on the inner side of the wall. For *Parthenocissus* the moisture content was very low (below 1% by weight) in both plant covered and bare brickwork.

Other Physical Damages

Generally, this study found very little damage caused by climbing plants

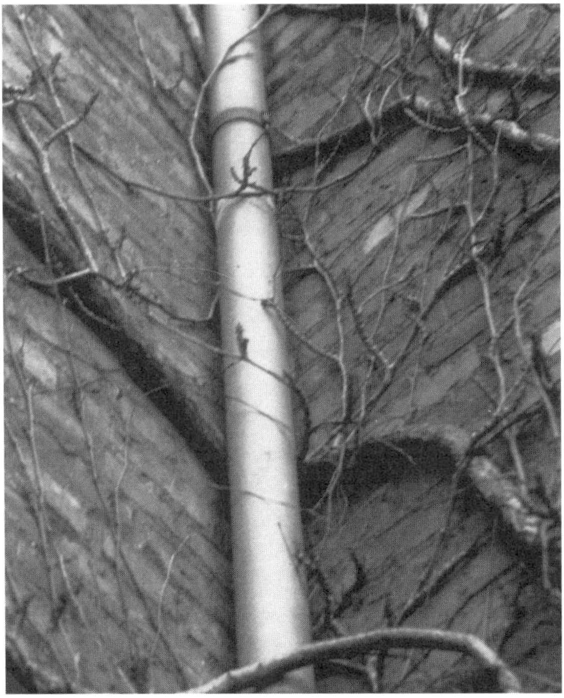

Figure 4. The Japanese Ivy develops thick main stems when it grows. These thick main stems can cause damage to downpipes by growing between the wall and the downpipes. If one plant is planted in each brick section seperated by downpipes and if the plant is kept away from the downpipes, problems can be avoided.

on walls. Most of the damages found could have been avoided through routine maintenance. For example shoots that fasten to parts of the building at acute angles to the front must be removed since the thin shoots might grow under them and crack the bricks. Likewise, the room between the brick front and downpipes must be kept to a minimum. The problem of lifting of tiles is exaggerated compared to reality.

Due to the efficiency of the fastening organs of the self-fastening climbers, paint and porous or defective plaster might be torn off when the shoots are removed from a front. It is often seen that the fastening organs leave visible marks on painted woodwork as well as on plastered and bare walls since they, to a certain extent, stay on the surface when the shoots are removed. These marks do blemish and are difficult to remove. However, it can hardly be called damage as it is quantitatively insignificant (2).

Energy and Indoor Climate Effects

A general assumption, especially with the evergreen climbing plants on walls, is that they have an insulating effect. The insulation layer is considered to be especially effective on non-insulated walls with a very dense evergreen plant cover. In Germany a number of temperature measurements have been carried out on buildings with climbings plants. The conclusion of these measurements is that the air under a plant cover of self fastening climbers acts as an insulation layer between the wall and the air outside the leaf layer. In general, with *Hedera* the wall is cooler during the day and warmer during the night and is cooler during the summer and warmer during the winter.

A dense plant cover of *Hedera* on outer walls in a European study shows an improvement of 33% on poorly insulated walls and 10% on walls with a normal insulation standard (14). Consequently, the insulation effect is positive, but small, for buildings with a normal insulation standard.

Temperature equalization over the day and the year is also important for the thermal tensions. Thermal tensions occur in brickwork as a result of matter expansion when temperature rises. In extreme cases, the expansion and contraction may result in cracks, especially in new brickwork with the risk of subsequent moisture damages (15).

In general, large quantities of plant cover may contribute to lowering the temperature since they do not get as warm as the surrounding air. Thereby they have a cooling effect without shading.

Urban Climate and Air Quality

There is no doubt that climbing plants on walls influence the urban climate and urban air quality. The urban temperature is decreased on hot days and the atmospheric humidity is higher at places with climbing plants on walls. In addition, the wind velocity is decreased due to the increased roughness of the building surface. Furthermore, like other plants, the climbing plants consume carbon dioxide and produce oxygen.

The occurrence of climbing plants on walls is still rare. In Berlin less than 1% of the building surfaces are covered with climbing plants, and therefore it is not possible to find a measurable effect on the urban climate. However, an effect on the microclimate in special sites can be found, e.g. in courtyards (10).

The large plantings contribute to cleaning the air. Measurements of dust and heavy metal uptake show that maximum 6 g/cm^2 can be deposited on *Hedera* and *Parthenocissus*. On *Parthenocissus* fine particles fasten on the leaves. However, the leaves must be removed in order not to leave the particles in the urban environment at time of decomposition.

In comparison with specifications for urban trees, a *Hedera* planting has approximately 12 m^2 leaf surface per m^2 and a big 100-year-old beech has 1,600 m^2 leaf surface on the approximate 170 m^2 projected canopy area (14). So climbing plants on walls can be just as effective as trees in respect to urban climate and air quality.

Climbing plants on walls probably reduce the noise by preventing reflections. However, this has not been verified in any research projects.

Aesthetic and Recreational Qualities

Climbing plants on walls have a great importance as food source and nesting site for birds. *Parthenocissus* and *Hedera* species are fruit–bearing when the plants are old enough to develop flower shoots. In addition, insects in the plants act as a food source for birds.

There are many good examples of architecture where climbing plants have been established as part of the architectural expression. In Denmark, there are examples where climbers are a part of the expression of the building and also function by working as a shadow curtain in the summer season. Considering its aesthetic, architectural as well as psychological qualities, climbing plants are important as part of the vegetation of urban areas where there is very little room for other vegetation.

Conclusion

Based on a examination of the literature and moisture measurements of Danish walls, it can be concluded that there are many advantages and only a few disadvantages for climbing plants. Moisture damages due to climbing plants can be rebutted.

The advantages of self fastening climbers in relation to building techniques are mainly insulation of walls and protection of walls against heavy acid rain and thermal tensions. However, the insulation effect is modified by the insulation condition of the building in general.

The physical disadvantages of climbing plants can all be prevented through a regular maintenance.

Acknowledgment

The project with the results discussed above is financed by the Danish Ministy of Housing. I wish to thank senior researcher Karen Attwell, Danish Building Research Institute who conducted a comprehensive literature study, Engineer Jens Østergaard Danish Technological Institute, Masonry Center, who tested the brickwork and participated in the study of damages and graduate Landscape architect student Hans Plovstrup who established and carried out the field study.

Literature Cited

1. Althaus, Christoph. 1987. Bauschäden durch Selbstklimmer. Gartenpraxis 8/1987, pp. 25–28.
2. Althaus, Christoph, Kiermeier Peter. 1989: Freising. Der Efeu als Fassadenpflanze – Nützlich, aber nicht risikofrei. Neue Landschaft 34/1989, pp. 533–538.
3. Althaus, Christoph. 1987. Bauwerk und Fassadenbegrünung mit Kletterpflanzen – Risiken, Schaden und präventive Schadensverhütung. Das Gartenamt 36/19–87, Mai.
4. Althaus, Christoph. 1987. Fassadenbegrünung. Ein Beitrag zu Risiken, Schäden und präventiver Schadensverhütung. – Patzer Verlag, Berlin–Hannover. ISBN 3–87617– 070–2.
5. Baumann, Rudi. 1980. Pflanzliche Verschattungselemente an der Gebäunde-oberfläche als Massnahme zur Reduzierung der Strahlungsbelastung unter sommerlichen Bedingungen. Inauguraldissertation, Gesamthochschule, Kassel. Druck: Karl Strube, Gensungen.
6. Bruhn, W. 1910. Beiträge zur experimentellen Morphologie usw. S. 98–164 in Flora/allgemeine Botanische Zeitung. – Vol. Neue Folge 1. Verlag von Gustav Fischer in Jena.
7. Bärtels, Andreas. 1987. Kletterpflanzen in der Stadt. Baumschulpraxis, 1987, v. 17 (3), pp. 93–99.
8. Emery, Malcolm. 1986. Promoting Nature in Cities and Towns. Ecological Parks Trust, U.K. ISBN 0–7099–0970–5.
9. Havens Planteleksikon. 1970/78. Trœer og Buske. Bind I og II. De Samvirkende Danske Haveselskaber (In Danish).
10 Köhler, M, og Bartfelder F. 1987.Stadtklimatische und lufthygienische Ent lastungseffekte durch Kletterpflanzen in hochbelasteten Innenstadtbezirken. Verhandlungen der Gesellschaft für Ökologie (Giessen 1986), Band XVI.
11. Ludvig, Karl og Trillitzsch Falk. 1983.Möglichkeiten der Fassadenbegrünung. S. 44–57 In Das begrünte Haus. Fundamente Alternativer Architektur. Band 10. Karlsruhe. ISBN 3–7880–7225–5.
12. Metcalf, C.R. og Chalk L. 1983. Anatomy of the Dicotyledones, Vol. II, Clarendon Press, UK, 1983.
13. Mikkelsen, Valdemar M. 1968. Botanik I. Morfologi. DSR forlag. Kbh.
14. Minke, Gernot og Witter Gottfried. 1983: Häuser mit Grünen Pelz. Ein Hand buch zur Hausbegrünung. – Verlag Dieter Friske GmbH, Frankfurt. ISBN 3–88184–062–1.
15. Siebel, L. 1973. Oberflächentemperaturen an Gebäudeaussenflächen. Deutsche Bauzeitschrift 9/1973, pp. 1757–1766.

A Critical Look at the Potential for Foundation Damage Caused by Tree Roots

Dealga O'Callaghan and Michael Lawson

Trees have been blamed for damage to the foundations of buildings in Britain. This has resulted in a large number of insurance claims. The approach to dealing with the problem is based on the concept of water demand of mature trees leading to remedies that involve severe reductions in size of trees on frequent cycles. The concept is shown to be fundamentally flawed and the legal position of the recommended practices in terms of liability in negligence is reviewed. The role of a subsidence risk assessment procedure is discussed.

For many years, trees in Britain have been implicated in damage to the foundations of low rise buildings (15). Damage is alleged to occur by the activity of roots which can dry clay soils, which can in turn cause subsidence. This has resulted in a large volume of claims against Insurance companies that underwrite the risk of subsidence and heave in buildings insurance policies. The increase both in the volume and magnitude of claims has been reported elsewhere (15) although more recent figures show a trend towards a slight reduction (Figure 1).

The problem is based on the concept that soils with a significant clay

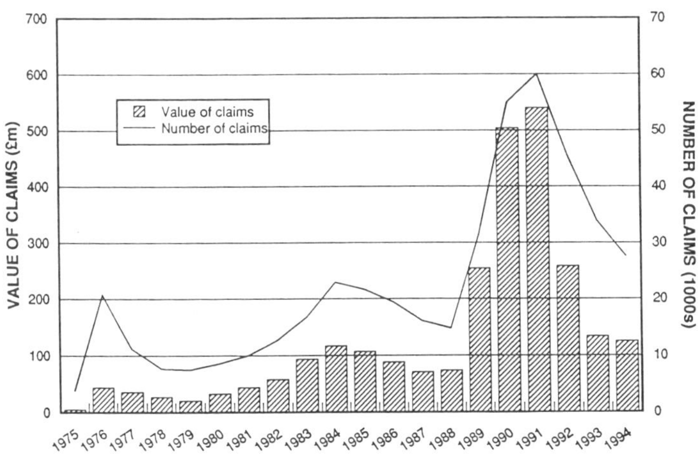

Figure 1. Annual value of insurance claims for subsidence and heave damage to housing. Reproduced by kind permission of the Institution of Structural Engineers.

O'Callaghan Associates Ltd, Valleyfield House, 1a Stratford Road, Aigburth, Liverpool, L19 3RE, England

content can change in volume with the addition or removal of water. If a building foundation is resting on a clay soil, the movements which result from changes in moisture content can, in some circumstances, result in damage to the building (15,17). Although the moisture content of soils is dependant upon the addition of water through precipitation and its removal due to a number of factors, including evapotranspiration by vegetation, it has been alleged that the most significant form of vegetation in this regard, and by inference the main causal agent of damage, is the tree (1,2,3,5,17).

An industry has evolved around the need to protect buildings from what is perceived to be the ravages of trees and their roots, in order to protect the financial investment of both owner and insurance carrier. There are two main aspects to this industry; first the need to deal adequately with the repair of buildings which have failed, and, second, the need to undertake a risk assessment of the likelihood of trees damaging buildings in the future. The former involves deriving solutions to an existing problem, hopefully, through sound knowledge and engineering principles. The latter, however, involves making a value judgement on the likelihood of damage occurring given the type of soils present, the species of tree, the climate and the type of building.

When buildings are threatened, or perceived to be, by the activity of tree roots, solutions are sought which deal with the problem on a one time basis. Both the law and the insurance industry prefer final and definite solutions. Such solutions usually involve underpining the foundations (17) and removal of the trees in order to attempt to "stabilise" the soil (2,3), or severe pruning to reduce the crown and thus, it is believed, water use by the tree (2,5,17).

The Water Demand of Trees

The published material dwells heavily upon the concept that different species of trees have different requirements for water depending upon their age and height. This has been termed "water demand" and has resulted in trees being classified according to their "demand" for water as high, medium or low (15,20). However, it is generally agreed that the water demand of trees is difficult to measure (15,17). Nevertheless, the concept that trees differ significantly in their demand for water persists among arborists and surveyors.

The term water demand has been defined by Lawson & O'Callaghan (15) as *The amount of water required by a tree in order to keep its metabolism functioning at optimum levels.* In this context it is difficult to see how tree species would differ significantly in this regard as physiological and metabolising organisms. Indeed, a review of the available literature would suggest that species differ very little in their water use as measured by transpiration rates, etc. (13). However, water demand is very difficult, if not impossible, to measure.

The concept of "water demand" is actually not accurate in biological terms according to Biddle (15), who coined the phrase (1) and was responsible for its inclusion in the National House Building Council Standards, Chapter 4.2,

Building near Trees (20). Biddle has confirmed that the concept should be taken to mean the *lateral extent, depth and intensity of soil drying which is achieved by different species.* In this context the concept of tree species having different rooting habits would seem to be a more rational explanation (15). Indeed the concept of differential rooting is generally accepted and has been classified to some degree by Binns (4).

Notwithstanding the above, McCombie (17) attempts to reclassify trees according to their water demand by adjusting the Kew Root Survey figures (10), and concludes that the adjusted figures represent an accurate way of predicting the likelihood of foundation damage as a result of the activity of trees (16,17). He makes the statement that it is rare in the United Kingdom for soil to become so dry that water is not freely available to a tree (17). Therefore trees are free to transpire as much water as their physiology will allow at all times, which seems to indicate that his view is that water is not a limiting factor for trees in the Britain.

Climatic data would tend to contradict this view (18). The Meteorological Office has been recording the soil moisture defecits (SMD), throughout the United Kingdom since 1960 and has constructed an extensive database on rainfall and evapotranspiration in relation to SMD. This is known as MORECS (Meteorological Office Rainfall and Evapotranspiration Recording System). This has been undertaken for agricultural purposes and basically refers to soils under grass or other vegetation, excluding trees.

Analyses of these data produce very interesting results (Figure 2). Two years results are illustrated here from a grid on the Cheshire Plain. It is accepted that generally there are upwards of 120 mm of available water in soils, therefore, when deficits approach this level, they are severe. Comparison between the years is interesting, It is apparent that in the year 1988 deficits were never extensive under grass and that field capacity was achieved within the year. However, during 1991, under the same field conditions, deficits were at maximum levels and there were no trees present in this study. However, field capacity was achieved in 1991 but the deficit persisted well into the winter.

It is clear that SMD occurs whether or not trees are present and therefore to base a predictive model for damage to foundations upon the relative water demand of trees, which is not an accurate parameter in any event however it is calculated or adjusted, does not seem to be logical.

Furthermore, McCombie asserts that the Kew Root Survey (10) and the NHBC Standards (20) are both indicators of the relative water demand of mature trees (17). Both of these assertions seem strange as Cutler has stated that the Kew Root Survey, of which he was one of the authors, is simplistic and has been misinterpreted over the years (9). Furthermore, Cutler states that there are significant anomalies between the planting frequency of the so-called high water demanding species and the frequency of claims associated with them (9).

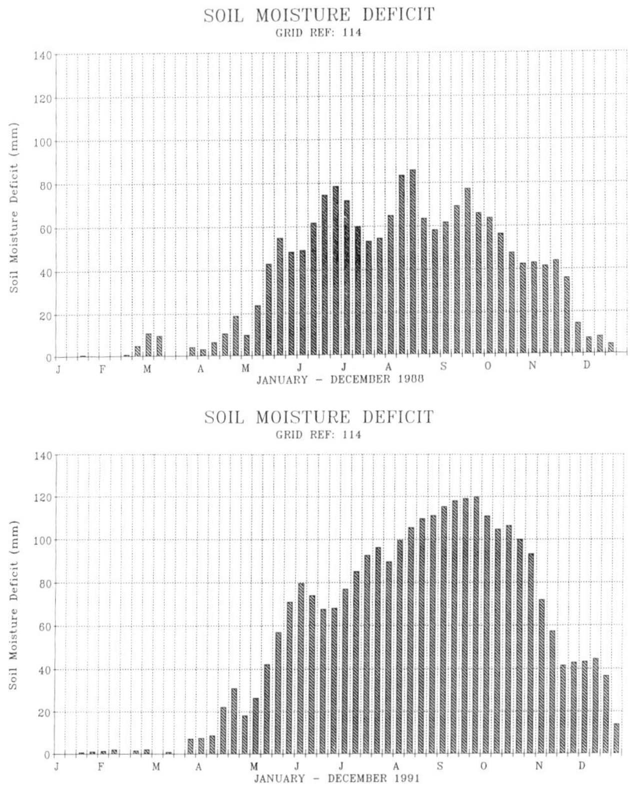

Figure 2. Table showing the soil moisture defecits on a month by month basis at one location in the Cheshire Plain during 1988 (A) and 1991 (B). Reproduced by kind permission of the Meteorological Office.

The NHBC have stated that their data have no bearing upon arboricultural matters; they are builders data produced for builders (Prince, pers. comm.). Furthermore the NHBC guidelines are for use in “new build” situations only and not designed to be extended to other situations; a point which seems to have been overlooked by McCombie (17).

The Current Practices

When buildings are damaged, it has been recommended that trees are either pruned to reduce leaf area or felled in order to stabilise the soil (2,17). However, pruning in this context is taken to mean crown reduction and not thinning or lifting, as reduction is thought to provide the greatest benefit (2). In reality, what happens is that the trees are severely reduced or more often topped out, rounded over or headed back.

This severe reduction or heading back has been reinforced by Biddle who states that only reductions of 50% or so are likely to have any effect on the water uptake of trees (3). However, he also states that the benefits of this are likely to last for one or possibly two seasons at best, after which the tree would need to be reduced again (3), as new growth and foliage increases water uptake by the tree. This view is reinforced by McCombie who advocates controlling trees by regular pruning (17). These statements have had the result that some City and Borough Councils have embarked upon a programme of 50% reductions of all their street trees on a two year cycle in order to reduce the risk of subsidence claims against them. Clear examples of this can be seen in some of the London Boroughs and in Bristol.

Notwithstanding the advice that only where trees are within a few meters of buildings are reductions likely to have any effect (2), the reductions continue within the London area and elsewhere. The consequences of such a policy are far reaching, in so far as the effects of the removal of 50% of the foliage of street trees may have upon the levels of particulate pollution and the "knock on" effect this may have upon human health in terms of increased respiratory disorders. Increased damage to the facias of buildings from particulates in the atmosphere can also be expected.

The Legal Framework

The policy of 50% reductions, which in effect amounts to topping, heading back, rounding over, etc., has interesting legal implications for the practitioners who undertake it and for those who institute the policy. Liability could also extend to those upon whose advice the policy of 50% reductions has been based.

English law has evolved over the last thousand years or so and is based on Statute Law and Precedent from Case Law (14). Over the years there have been many precedents in many aspects of law, but one significant case in the matter of professional negligence is the case of *Bolam-v-Friern Hospital Management Committee, [1957] 2. All E.R. 118 in* which Judge McNair stated, in the case which involved medical negligence (pages 425-426 of 14)

> *A doctor is not guilty of negligence if he has acted in accordance with a practice accepted as proper by a reasonable body of medical men skilled in that particular art. It does not matter that there is a body of opinion who take a contrary view.*

Significantly he went on to state that:

> *This statement is, of course, equally applicable to other professions......and that the most practical suggestion for professionals (in order to avoid negligence claims) is to follow strictly the recommendations of their professional bodies, for example financial reporting standards & other published material.*

Based on the above and, as prudent consulting arboriculturists, if we are to look for protection from negligence in the decision of McNair, then we should be following the published guidelines of our professional bodies. In the United Kingdom this would be the Arboricultural Association (AA), while in the United States it would be either the International Society of Arboriculture (ISA), or the National Arborists Association (NAA). Consultation with the published guidelines on pruning by the above (7,11,12), and the *British Standard Recommendation for Tree Works BS 3998:1989* (6), all advise against topping, heading or severe reductions. A glance through the world-wide literature on pruning tends to support the view that topping, etc., is bad practice (11,13,21,22). The hazard potential of topped trees is also recognised throughout the literature (6, 7,11).

It would seem therefore that the practice of severe reductions to reduce a perceived risk of subsidence is contrary to the accepted practices within the arboricultural industry. It is also significant that in the United States, the California Legislature recently enacted a law which effectively outlaws topping, i.e., *Government Code Section 53067*, which is justified by the following narrative:

> *Topping is the practice of cutting back large diameter branches of a mature tree to stubs and is a particularly destructive pruning practice. It is stressful to mature trees and may result in reduced vigour, decline and even death of trees. In addition branches are weakly attached to the tree and are in danger of splitting out.....*

Whether or not this will ever be enforced remains to be seen. However, in the United Kingdom, negligence claims are increasing and professionals are particularly vulnerable. Recommendations that advocate such large reductions in canopy to lower a possible subsidence risk are likely to be questioned. Particularly as the larger and older trees decline and die following the continued reductions.

Discussion

The relationship between trees, roots and soils is a complex one. Roots do abstract water from the soil and tree roots can dry a full horizon quite easily. However, tree roots are also capable of absorbing water at depth and giving it up to the soil at shallower depths (8), a phenomenon known as hydraulic lift. It has also been suggested that with the distance of root water withdrawal influence at high diffusivity, that root extension can outstrip the drying effects to reach untapped moisture (8). In short, the whole subject area is very complex and relatively little is known about it.

There does not seem to be any simple or linear relationship between the perceived water demand of different tree species, their planting frequency and damage claims (9). There is, however, a direct relationship between the incidences of claims and increases in SMD (19). The relationship is illustrated in Figure 3 where the number of claims has been plotted with the SMD from 1976 through to 1990. It is important to remember that this is based on the MORECS data and there are no trees involved. When this hypothesis is tested statistically by plotting the actual number of claims recorded against

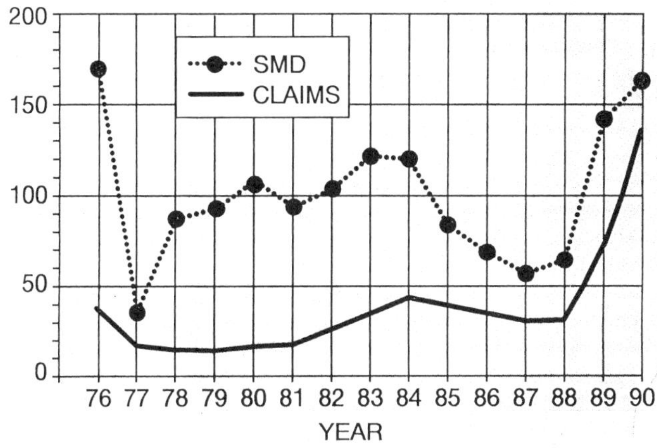

Figure 3. A plot of the occurrence of claims for subsidence in relation to recorded soil moisture defecits (SMD) from 1976 to 1990. Note the relationship of the claims to the SMD. Reproduced by kind permission of the Meteorological Office.

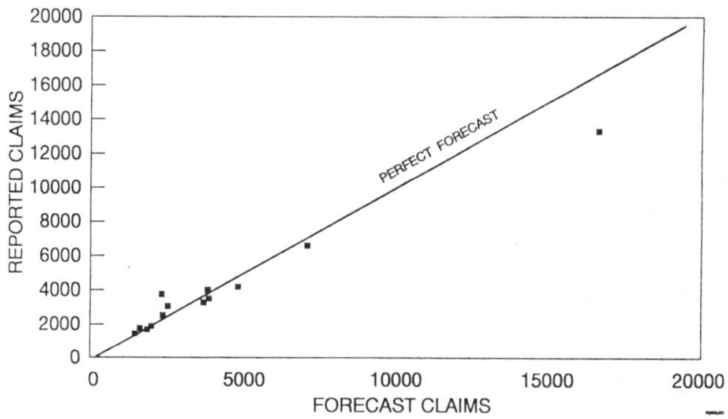

Figure 4. A test of the hypothesis of the actual number of claims against the forecast level of claims. Note that the data are a very close fit to the perfect forecast. Reproduced by kind permission of the Meteorological Office.

the forecast of claims, a linear relationship is the result (Figure 4). Again there are no trees involved.

The relationships become much more complex when trees are introduced to the system and when it is acknowledged that most of the claims are in urban areas where the soils are far from optimal for tree growth. The roots of trees in these environments are not able to exhibit the form and extend to the same degree as those in a natural or wild environment (8). Other factors which influence the system include genetic variability not only between genera and species but between individual trees of the same species. The challenge is to interpret what is happening in the face of individual site conditions, complex and often altered soils, the variability of the climate and the known profile of individual tree species.

The use of the term "water demand" as a basis for assessing the probability or likelihood of trees to cause damage to building foundations is fundamentally flawed. Persistence in the use of this approach will damage the urban tree population beyond repair. It mitigates against the maintenance or planting of trees in built up areas and thus affects the quality of all our lives. The concept has already caused local authorities to react by reducing the crowns of their street trees by 50% on a two-year cycle because of a perceived threat from the use of water by the trees.

It is important that, in planning urban tree strategies, all aspects of the urban tree population be considered. The benefits and positive effects should be balanced against the costs and threats, including subsidence risk. Trees should be assessed for hazard to protect local authorities from claims should collapse occur. Is it not time that "subsidence risk" was addressed in similar fashion? A Local Authority should take whatever steps are prudent to cover their liability. Swift and "knee jerk" reactions based on little knowledge and fact is, with respect, not a prudent approach.

What is needed is a dialogue with engineers and building professionals to design foundations that are capable of withstanding the level of soil movements experienced in Britain. As arboricultural professionals, we must supply our colleagues in the building, insurance and legal professions with factual and unambiguous data about trees. It is up to them to design the solutions.

Acknowledgments

We thank the many people who have assisted us throughout this work. In particular we are grateful to Mike Crilly of the Building Research Establishment, John Mead of Zurich Municipal Insurance Company, Frazer Broomby of Provincial Insurance Plc and Brian Clancy. The continued assistance and interest of Stewart Wass of the Metereological Office is greatly appreciated.

Literature Cited

1. Biddle, P.G. 1983. Patterns of soil drying & moisture deficit in the vicinity of trees on clay soils. Geotechnique 83 (2): 107-126.
2. Biddle, P.G. 1992. Tree roots & foundations. Arboricultural Research Note 108/92/SILS. 6pp Arb. Advisory & information Service, Alice Holt, Farnham, England.
3. Biddle, P.G. 1995. Tree roots: indirect action. In Patch, D (Ed) Proc. of Arboricultural Practice: Present & Future Conference, Warwick, England (In Press).
4. Binns, O. 1880. Trees and water. Forestry Commission Arboricultural Leaflet No.6, 19pp, HMSO Publications, London.
5. Borden, J.M. & R.M.C. Driscoll. 1987. House foundations - a review of the effect of clay soil volume change on design performance. Mun. Engr. 4: 181-213.

6. British Standard 3998. 1989. Recommendations for tree work, BSI Publications, London
7. Capel, J. & D.H. Thorman. 1994. (2nd Ed)Tree pruning guidelines. Arboricultural Association, Romsey, England, 40pp.
8. Craul, P.J. 1992. Urban Soil in Landscape Design. John Wiley & Son Inc. 396pp.
9. Cutler, D.F. 1995. Interactions of tree roots & buildings. In Watson, G. & D.Neely (Eds) Proc of Trees & Buildings Conference, Lisle, Illinois, ISA Publications (In Press).
10. Cutler, D.F. and I.B.K. Richardson. 1989. (2nd ed). Tree Roots & Buildings, Longmans, Essex, UK. 71pp
11. Harris, R.W. 1992. (2nd ed) Arboriculture: Integrated Management of Landscape Trees, Shrubs and Vines. Prentice & Hall, Englewood Cliffs, NJ 674pp.
12. Hibberd, B.G. 1989. Urban Forestry Practice. Forestry Commission Handbook No.5. HMSO Publications London, 150pp
13. ISA 1995. Tree Pruning Guidelines. ISA Publications, P.O. Box GG, Savoy, Illinois, 14pp
14. Keenan, D. 1995. (11th ed). Smith & Keenan English Law. Pitman Publications, London, 820pp
15. Lawson, M. & D.P. O'Callaghan. 1995. A critical analysis of the role of trees in damage to low rise buildings. J..Arboric. 21: 90-97.
16. McCombie, P.F. 1993. The relative water demand of broad leafed trees-A new analysis of the Kew root survey. Arboric.J. 17: 359-374.
17. McCombie, P.F. 1995. The prediction of building foundation damage arising from water demand of trees. Arboric. J. 19: 147-159.
18. Meteorological Office. 1993. Internal Report on Rainfall & Evapotranspiration Records. S.Wass to M.Lawson, Personal Communication.
19. Meteorological Office. 1994. Internal Report on Soil Moisture Defecits and their Relationship to Insurance Claims for subsidence. S.Wass to M. Lawson, Personal Communication.
20. National House Building Council (NHBC).1992. NHBC Standards Chapter 4.2 Building Near Trees. NHBC Buildmark House, Amersham, Bucks HP6 5AP.
21. Shigo, A.L. 1989. Tree Pruning - A Worldwide Photoguide. Shigo & Trees Associates, Durham, NH. 186pp.
22. Shigo, A.L. 1991. Modern Arboriculture. Shigo & Trees, Associates, Durham, NH. 424pp.

PART III

Tree Protection and Preservation

Tree Quality BMPs for Developing Wooded Areas and Protecting Residual Trees

Kim D. Coder

Key components of managing trees and forests on development sites rest in tree-literate site evaluation and damage control activities. Tree and site attributes and their relationship with the design process and construction methods determine post-development survival, continued tree success, and the requirement for treatments. This paper reviews the three tree quality management time periods on a development site. Best Management Practices (BMPs) are defined that have proven to provide resource owners with positive tree preservation results over the long-run. Pre-development evaluations, development site management, and post-development care are areas where current and future tree quality concerns are discussed.

A house, business, or green space shaded by trees is a desire of many people. Trees add value to a development site by generating services or functions that protect us from the elements, shield us from the harshness of the human world, surround us in a romanticized "natural" environment, and provide us with physical and psychological comfort. Trees also increase real estate values in many situations. Coupled with the great demand for trees and its associated values is a lack of tree-literate understanding of real sites and real trees.

Our role in arboriculture and community forestry is to guide people and resources to meet specific goals and objectives following a management plan. To assist in this process, protection of tree and site quality must be key aspects in developing a site. Stated simply, to assure whole tree quality, all tree parts must be protected from acute and chronic damage, and short and long-term site resource degradation must be minimized.

Best Management Practices

Managing tree quality on a site through the use of proven **B**est **M**anagement **P**ractices (BMPs) is critical in attaining the three aspects of tree-literate stewardship: 1) tree and site asset protection; 2) appreciation in asset value and sustainability of asset functions over time; and, 3) minimizing liability risks. A tree quality management program concentrating on these three aspects is essential around development sites. An extensive educational program has been successful in changing behaviors concerning trees and development (1,2,5,6).

Associate Professor / Extension Forester, Extension Forest Resources Unit, School of Forest Resources, University of Georgia, Athens, Georgia 30602

A tree quality management program helps to build awareness of tree and forest attributes, functions, and values among planners, builders, and buyers. In areas where development is rampant and fast-paced, insertion of best management practices (or BMPs) for trees, targeted at specific audiences, can be successful. These BMPs can be in the form of voluntary guidelines, regulatory edicts, or educational outlines. Whatever the form BMPs take, both the supply side and demand side of development activities must be reached.

Tree quality BMPs can be grouped into three distinct time classes: A) pre-development and site evaluation; B) development activities; and, C) post-development damage mitigation and monitoring. Pre-development BMPs revolve around site selection, project planning, and tree and forest attributes. Development BMPs concentrate on site-layout, tree protection zones and site-damage control. Post-development BMPs concentrate on restoration of tree functions and values.

All three classes of tree quality BMPs attempt to minimize development problems while protecting tree quality by following two purposes: 1) maximize a tree's biological efficiency in reacting to environmental changes; and, 2) minimize potential damage to a tree and its site. There are a number of activities and cultural treatments that are effective in reaching these twin purposes.

Tree quality BMPs must be interwoven early in the design process and continually with construction methodologies. Each site and tree have a unique palate of resource attributes, opportunities and constraints. The use of BMPs must be accepting of actual site attributes as well as the desires and perceptions of tree or site owners.

Client perceptions. Examining client expectations and desires are essential for tree quality and professional success. Every development situation is different partially because of client concepts and expectations. Clearly understanding these expectations will allow for meeting them more fully and prevent misunderstandings and post-development regrets.

Development sites can range from having no trees to being covered with a dense tangle of tree forms. Every person's perception of function and value will differ for the same site. To many clients, tree quality expectations revolve around development having a neutral or positive impact on the site resources over the short and long-run.

Many people do not — can not — appreciate what the act of building a structure or changing land use will have on the natural resources currently present on a site. Trees on development sites are expected to remain unblemished and as vigorous as before. Site constraints are expected to diminish. Forests are expected to regenerate and grow. Expectations of ideal results, regardless of site and tree conditions or development activities, are marketed to clients but represent fantasy. Education remains a key component of our jobs around a development site.

An examination of human perceptions on wooded development sites concern what a forest should look like and include: domination by young to middle-aged, even-aged trees; medium to large trunk sizes; wide spacing between trees; open-grown crowns; evenly distributed, not clumped; open under- and mid-story sight lines; small pockets of understory "brush"; a diversity of common plants and animals; natural noises and smells; and no clearly human-initiated damage (partially derived from citation 8).

Pre-Development Evaluation

Given these perceptions, how do we develop a wooded site and maintain tree quality? Pre-development tree quality BMPs require site and tree evaluations. Features to examine include: the amount of space available; tree species, size, number, form, and condition; soil characters; past damage; and expected responses to change.

Tree quality management around development sites must begin early in the planning stage as part of a site-team effort. It would be best to begin the pre-development evaluation process at least one growing season in advance of site development activities. Table 1 lists the major features examined in a pre-development examination.

One of the first features to examine is building lot size. It is difficult to generate forest-like functions and values on lots smaller than one acre for average sized houses. The common one-quarter to one-third acre house lots have little room on which to develop a wooded character. Here individual trees and tree clumps must provide expected functions and values.

For business development sites with premium per square foot values and specific land-use requirements, like parking places, a wooded atmosphere is difficult to generate without significant area dedicated to linear sight - breaks, and clump zones. Trees generate many functions and values on a development site but are not 'free' or without costs. Tree quality protection and tree planting do remain a highly cost effective addition to a development site.

Tree-site attributes. Another feature to consider in evaluating a site is the number of trees. As a rule the more trees on a site, the more a forest-like atmosphere is generated up to a point. Over-abundance can be a problem.

Table 1. Major features of a pre-development examination.

Lot Size / Space	High Grading
Number of Trees	Planting / Moving / Saving Trees
Tree Density	Soil Surface
Tree Species	Soil Quality
Tree Sizes	Soil Slope and Depth
Tree Crowns	Biology-First Design
Past Damage	Treatments
Establishment of a Tree and Site Management Program	

Sites with too many trees can be as limiting and unresponsive to development as sites with too few trees. Building in dense, over-stocked stands result in decline and death of many trees. Remaining trees are prone to windthrow, pest, storm and ice damage.

Determine the biological occupancy level of trees on a site. Any site, depending upon its inherent productivity, can only hold a given amount of leaf surface area. This leaf surface area can be concentrated on a few large trees or on many small trees. For example, 600 five-inch trees, 150 ten-inch trees, 65 fifteen-inch trees, or 40 twenty-inch trees all carry similar total crown volumes per acre. There is a trade-off between the numbers of trees and their sizes for similar site occupancy (Fig.1).

One way to estimate site occupancy or tree density is by measuring basal area (BA). Basal area is a forestry measure that determines the cross-sectional area (in square feet) of all the trees on an acre at 4 1/2 feet above the ground. This measure is made using an angle gauge or prism at a number of random points in an area. Basal area is an approximation for crown area or site occupancy.

Using basal area can help quickly establish site occupancy, and expectations of how a site will respond to development. A site would be considered over-stocked and unresponsive if basal areas are greater than 70 square feet per acre. A site is under-stocked if basal areas are less than 35 square feet per acre.

Another site feature to examine is the tree species mix. There are many types of trees and associated reactions to site changes. Some trees will respond well to nearby construction and associated environmental changes, others will begin to decline almost immediately.

The greater the variety of good trees on a site, the better the chances for the long-term health of all trees. A mix of several different species can insure

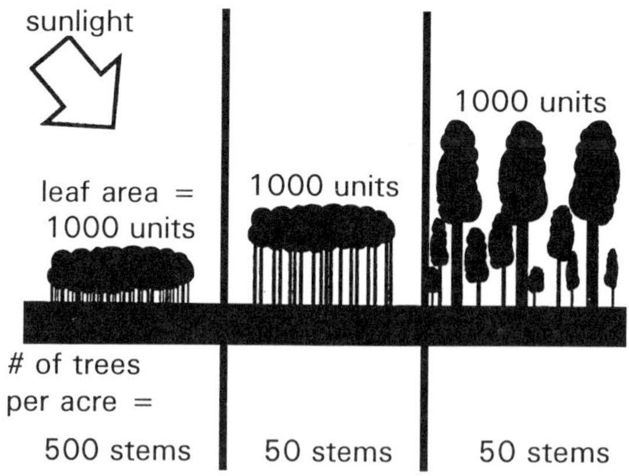

Fig. 1. Depending upon inherent site productivity, only a given amount of leaf area can be maintained. Crowns must be managed to meet an owner's tree cover objectives.

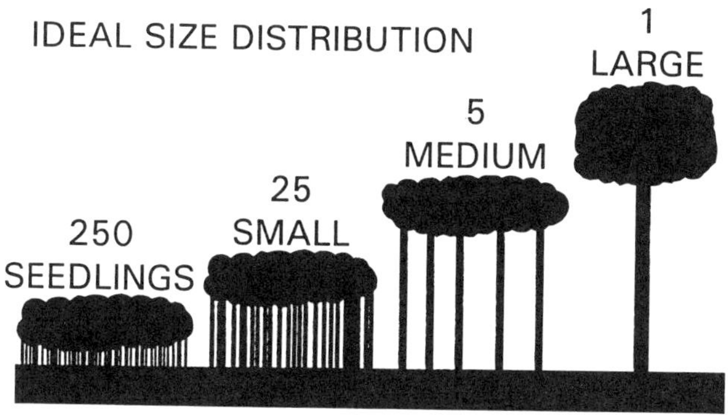

Fig. 2. The relative proportion of each size class of tree present affects the short-term and long-term aesthetic and biological quality of a site.

a healthy diversity of tree cover. Work for at least three species per crown class. If a site has only one species, such as only pine, only hickory, or only cottonwood, the site is at risk for tree problems. Invest in, and manage toward, genetic diversity.

A diversity of tree sizes is also important. There should be a mix of small, medium, and large trees. As discussed previously, client perceptions of wooded or forested sites revolve around tree attributes. Management of tree size ratios on a site maintains the forest-like perceptions. As a general rule, for every existing large tree (11 inches diameter or greater), there should be five medium sized trees. For every medium sized tree (5-10 inches diameter), there should be five small trees. For every small tree (1-4 inches diameter) there should be ten seedlings / saplings (Fig. 2). Adjust sizes from the large trees downward.

Across most sites, the more variety of tree sizes and species maintained, the more a forest-like perception will result. Thinning existing trees, removal of understory trees, and planting can help preserve or enhance these perceptions.

Another valuable evaluation technique is examining tree crowns as an estimate of a tree's response to change. The proportion of living branches or living crown should comprise at least 35% of mature tree height. Mature trees with a large volume of living crown can react most favorably to development. If the live crown is less than 25% of total height, problems may develop after construction. Small crowned trees (<15% live crown) would be candidates for removal before construction begins (Fig. 3).

Historic context. Past damage on a site must be recognized and mitigated, or worked around. Site and tree damage may have resulted from past construction, logging, storm, erosion, or land clearing. It is difficult to use machinery between and around trees without damaging tree parts or site

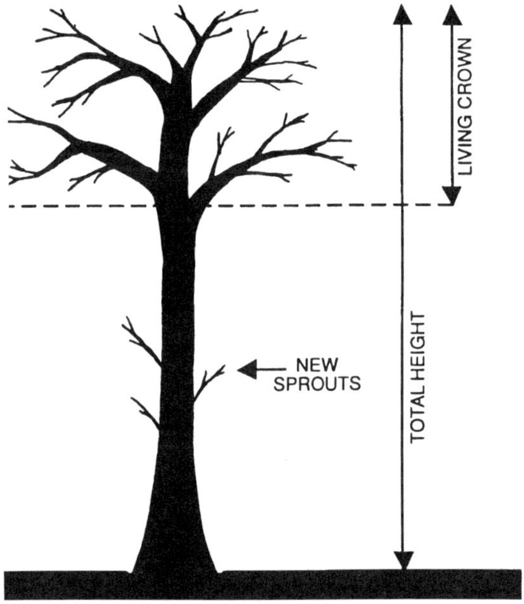

Fig. 3. In mature trees, the proportion of living crown to total tree height can help determine which trees are poor risks for site development. Do not count new sprouts along the trunk as part of the living crown.

resources. Such injuries lead to decline, decay, pests, susceptibility to additional damage, and structural problems. Severe mechanical damage will make it more difficult for trees to adjust to any site changes.

Carefully examine the bottom 20 feet on tree trunks and the basal 10 feet of root area for any scrapings, tears, or wounds. If mechanical injury disrupts more than 1/3 of the trunk's circumference, removal is warranted. The basal portion of the tree withstands the most concentrated structural loads and is most prone to debilitating damage. Prognosis for severe damage in this basal portion of a tree is usually poor over the long-run. Hazard assessments should be completed and hazardous trees removed before construction begins.

High-grading. One form of site damage not always appreciated is high-grading. High-grading is a forestry term which denotes that the best trees have been removed, leaving only declining, crooked, pest-prone and poor species of trees to reproduce and grow. Developing a site with poor quality, poor condition, and misshapened trees, of undesirable tree species, may lead to poor tree reactions to site changes and to long-term tree problems. Recognition of high-grading and designed enrichment plantings can minimize function and value losses.

Inter-plantings under moderate or dense canopy covers, and with no low level competition control, are rarely successful. An area where the new trees have room to effectively colonize above and below ground is important. Sunlight quantity and quality must be managed to insure planting success. Also, examine the site carefully and mark and move trees in non-tree areas. Root pruning by hand or with a tree spade, then later removing trees to a safe holding area, will greatly reduce landscaping costs and conserve local genetic material.

Soil quality. Another critical item for evaluation is the soil surface. The soil surface mirrors past site abuse and current health. The soil surface can show soil disruption, heavy equipment use, and compaction on the site. Soil movement across the surface from natural processes or from equipment can lead to tree damage and site productivity losses. Removing soil can severely damage roots. Excessive fill (defined primarily by soil texture) around existing trees can suffocate roots and cause tree decline or death.

Erosion from past and present activities can destroy site productivity and tree quality. The presence of many exposed surface roots and the lack of natural litter suggest excessive erosion, compaction, and drainage problems. Establishing tree protection areas well before construction is important. Physically protect roots and rooting areas with mulch, fences, plywood sheets, and other physical barriers.

The attributes of soil that allow a tree to survive and thrive must be defended. Soil moisture, aeration, organic matter, texture and structure should all be examined for constraints and potential cultural corrections. Based upon soil genesis factors, less slope and greater soil depth with good drainage, help generate tree quality and strong tree reactions to site changes.

Biology-first design. A good landscape design includes tree-literate features and tree-literate individuals involved in the process from the beginning. Design errors and planning faults remain a primary cause of tree decline and death. Anyone who has touched a tree may be blamed for negative consequences around a development site. But, often it is the initial, tree-illiterate design that directly caused tree quality loss. To alleviate many commonly occurring design faults with trees, a new design paradigm must be accepted. This concept is termed 'biology-first design.'

Major components of biology-first design include: A) do not force trees into aesthetic designs as structural shapes, textures, colors, or forms; B) never force biology, microclimate and soil attributes to meet design objectives and criteria; C) demand and be willing to pay for the best quality trees; D) identify site characters of biological importance and match them with real tree attributes; E) design a program of health and structural maintenance for trees as part of any planning; F) provide enough physical space for a tree to sustain and support itself over its effective life.

Intelligently plan for site circulation, sidewalks, driveways, parking spaces, turn-around areas, recreation areas, patios and decks. Each of these items requires a specific amount of space. Good planning and proper layout can truly minimize impacts on tree quality and site resources. This process requires tree-literacy and more than the tradition lip- or pencil-service usually given tree quality problems.

Pre-development treatments. After pre-development site and tree evaluations have been completed, treatments can be recommended to minimize damage and maximize tree reactions to change. Preferably one full growing season in advance of any development activity on the site, and if warranted by tree and site characteristics, treatments could include: marking utility and access corridors and assuring needed clearances; marking construction danger zones and tree protection zones; pruning, cleaning, dead-wooding, and clearing; mulching tree protection areas; setting-up tree protection barriers; establishing irrigation needs and methods; fertilizing with potassium, phosphorus, calcium, magnesium, and macro- and micro-essential elements if needed; and, making small nitrogen applications.

As you are aware, seldom are tree professionals allowed the luxury of timely advanced access to sites for evaluations or treatments. Usually tree professionals are called to fix tree quality concerns after site development activities have commenced and some construction damage has already occurred.

Development BMPs

The next class of tree quality BMPs to discuss is used when construction begins. The first and most critical rule in working with tree quality on development sites is to "get there first!" The first approach to a tree can be the most damaging and facilitate further damage. The tree quality manager needs to be the first person to approach all the trees on a site and make removal, treatment, and preservation decisions.

In addition, tree quality managers need to insure that every person on a site knows where the trees actually are — not some general circle on a site plan. Plans should include accurate and precise locations for the trunk, crown, and major soil areas colonized by the roots. Remember that construction and development, in themselves, do not damage tree quality and site resources — individuals accomplish these actions. Using on-site education, daily monitoring, and/or a strong series of fines, penalties, and rewards will help people to remember that trees are important.

Once a site plan is available, determine where tree quality and site resource damage are most likely to be concentrated. When all tree parts are not accurately recorded, take the site plans, a 100 feet tape measure, flagging tape, paint, compass, string, and wooden stakes to the site. Outline on the ground with a string where various planned structures and areas will occur.

Preservation and construction zones. If a tree is within thirty feet of the string, it is in the construction danger zone and should normally be removed to facilitate good construction. Trees within this zone are easily and consistently damaged during construction. These trees decline and die due to damage, or eventually become a hazard to structures and people in the area (Fig.4).

High quality trees between zero and sixty feet from any structure should be individually protected with barriers and stem, branch and root padding or wraps. The area between zero and sixty feet from structures is the tree protection zone. Trees already in poor condition should be removed. Use barriers to minimize damage to trees in the tree protection zone. They should be installed before construction begins anywhere on the site. Barriers will not prevent all damage but will remind people working on the site that trees are important and barriers should be respected.

Protect as much open soil surface as possible below the tree's crown. Trees require physical space and soil volume to colonize and control. Soil features needed include: A) physical space for support and pore volume; B) open surface area for oxygen and water movement; and, C) an adequate and sustainable supply of essential soil resources. To summarize this point, trees with large areas to grow in have the best chance of being healthy, and long-lived, and developing few problems. A tree quality manager assists a tree to colonize and effectively utilize a site.

Preservation space. There are a number of ways to determine how much space is minimally needed for tree survival and growth, and therefore the space that should be protected from development impacts. One type of measurement is determined from tree trunk size. Measure the diameter of the tree at four and one-half feet above the ground. For every inch of tree diameter, allow at least one foot to one-and-one-half feet of space away from the trunk (radius distance). A minimum root protection distance should always be at least six feet regardless of trunk diameter.

A second measure of a tree protection area examines the length of branches that extend horizontally farthest from the tree. This drip-line distance in feet, multiplied by 1.5 yields the number of feet away from the trunk to be protected. This measure requires careful application and may underestimate protection distances when trees have developed in dense stands. An absolute minimum of two-thirds the drip line distance should be defended. An effective hybrid determination is selecting the smallest or average measure of "drip-line X 1.5" and the previously mentioned "trunk diameter-inch" system.

A more effective means of determining a protection distance is using a site-occupancy measure. This method does not determine a minimal distance but an effective root colonization distance. The expected diameter (dbh) of a tree 10 years in the future is estimated. The expected diameter in inches is then multiplied by 2.25 to yield the number of feet (radius) of tree

Fig. 4. Delineating the construction danger zone and the tree protection zone assists with tree removal planning and tree protection measures.

protection distance. This system is based upon standard site occupancy estimates of 60 square feet of basal area and 750 square feet of soil area per foot of cross sectional area. Table 2 provides additional information using site-occupancy estimates for multiple island or clump trees (3,10).

Some construction damage will occur affecting tree quality and site resources. Construction damage is difficult to treat and tree professionals have a limited cultural toolbox with which to respond. Construction damage allows us little or no time, space, resources, or stored carbon to work with. If a tree is severely injured during construction, remove it immediately before the site is finished. Do not allow a badly damaged tree to become a pest center.

Always limit access of construction machines, material storage, chemical or cement rinsing, vehicle parking and site-office locations to non-tree areas. Do not let construction equipment near trunks or main rooting areas. Construction activities should not occur beyond 60 feet from the structures.

Soil problems. Soil level changes over the site can disrupt and destroy roots and negatively modify the soil environment. Fill and cuts, leveling, and surface cultivation or tilling all can damage or kill trees. An often overlooked

Table 2. Soil area overlap values per tree based upon site-occupancy values (2.25 X diameter inches) for use where trees share soil space in linear, island or clump plantings. (After concepts developed by Perry, 1990.)

Number of equal size trees neighboring tree A	Allowed area overlap values with tree A
2	40%
3	30%
4	20%
5	10%
>6	0%

but critical soil component is water availability and water movement. Soil changes or movement on a site can completely change water flow patterns, ponding, and soil aeration patterns. Soil cuts can drop water tables and available water away from established root systems.

Construction activities destroy soil resources, functions and values. Soil bulk density or compaction changes can be the most constraining and damaging. Compacted sites do not support vigorous tree growth and construction sites can easily have 50% greater bulk density than native soils. Increasing bulk density by one-third can be expected to cost a tree one-half of its root and shoot growth. (Note excellent review by Craul, 7.)

Soil compaction is a physical barrier to root growth and to oxygen movement. Tree roots have difficulty physically penetrating soils with bulk densities greater than 1.7 g/cc, and when the proportion of air pore space (macro-pores) in a soil drops below 15%. Table 3 defines more specific root limiting bulk densities by soil texture class (9).

Solutions to compaction problems on development sites include: A) deep tillage or sub-soiling; B) mulching; C) amending with large, porus, non-compactable solids to create an aerated soil framework; D) selective use of porus or open structure surfaces as long as compaction is not used to stabilize the units; F) deep core aerators (12-18" depth); G) vertical mulching with washed, graded, and non-compactable materials with the column tops open to the atmosphere (7).

A summary of tree quality BMPs as related to soils include A) prevention and restoration of high bulk density soils; B) avoidance, treatment, and prevention of soil contamination by construction materials; and C) improvement of nutrient cycling, moisture balance, and soil structure by addition of organic matter.

Site access. Design and control access corridors for utility installation, both underground and overhead. Depending upon local codes for underground utility corridors, two or three trenches will suffice for all the various utilities. Unfortunately, seldom do various utility providers cooperate in the

Table 3. Soil attributes, by soil texture class, where root growth becomes significantly limiting. (After Morris and Lowery, 1988)

Soil texture	Root-limiting bulk density (g/cc)	Root-limiting % pores normally filled with air
sand	1.8	24%
fine sand	1.75	21
sandy loam	1.7	19
fine sandy loam	1.65	15
loam	1.55	14
silt loam	1.45	17
clay loam	1.5	11
clay	1.4	13

installation processes to minimize tree quality loss. Utilities should be designated to non-tree areas for access, such as along driveways and sidewalks. Working with utility service providers to generate serpentine corridors and tunnels under tree rooting areas is essential.

Provide room for trucks and construction equipment to get back and forth to the building site. Two access points are needed because large equipment or delivery trucks cannot turn around without extensive site quality losses. An incoming and outgoing access should be designated for deliveries. This does require a designated, non-tree area for storage of construction materials and parking of construction related vehicles, including parking for laborers and sub-contractors.

Damage management. As development activities continue on a site, continually monitor damage to tree quality and site resources. Develop a damage class assessment to quantify managerial responses and to project expected tree life-spans and losses. Expected tree reactions to construction damage vary from immediate and out-right death, single-year decline and death, multiple-year decline and death, and decline with major amounts of living mass loss. The latter two are the most common results among residual trees, and the most difficult to prove cause and effect relationships with construction activities.

One way of summarizing construction damage into a number of classes is given in Table 4 in order of decreasing severity and likelihood.

There are a number of tree quality BMPs for development sites that can be gleaned from practice. These are included in Table 5.

Post-Development BMPs

Post-development tree quality BMPs primarily concern identification of problems and associated treatments that do not accentuate tree quality problems or further destroy site resources. In addition, sorting out the living trees from the dead, and soon to be dead, is required. Severely damaged

Table 4. Construction damage classes in order of decreasing severity and likelihood.

Class 1- General root system destruction
Class 2- Root collar and basal roots damage
Class 3- Mechanical damage to stem
- A) bark disruption
- B) chemical or fire damage
- C) wood damage
- D) signage damage

Class 4- Soil surface problems
- A) erosion / water availability
- B) litter loss / crusting
- C) compaction / fills / cuts
- D) temporary soil or material storage
- E) micro-climate changes

Class 5*- Wind load changes
Class 6- Major branch base damage

* Note that Class 5 above affects edge or island trees where clearing or thinning has left trees prone to windthrow. This is the only damage class not necessarily a result of direct mechanical or soil damage. (4)

Table 5. List of tree quality BMPs for development sites.

1) Know the development and building regulations concerning trees in your area.
2) Plan for a cement wash-out pit and designate a chemical holding area, both away from tree protection areas.
3) Limit site parking and storage to already damaged areas.
4) Allow no site-offices, equipment, or material storage in tree protection areas.
5) Control and limit on-site soil storage.
6) Control and minimize grade changes within tree protection areas.
7) Keep refueling and maintenance areas away from trees and native soils.
8) Control toilet, lunch, break and burn areas, and associated refuse.
9) Allow only two construction access drives to the site — one in, one out.
10) Control utility over-head and under-ground corridors.
11) Be careful of fire dangers to the site and surrounding woods, both during construction and afterwards.
12) Use tree protection barriers, wraps, and pads, keeping them in good repair.
13) Include contractual penalties in real dollars for tree protection area violations and tree barrier damage. Allow dollars to educate reluctant or tree-illiterate people and companies.

trees should be quickly removed and replaced with plantings, if warranted.

Cultural treatments on the post-development site can include: A) weekly water management (the most important item!); B) fertilizing with potassium, phosphorus, calcium, magnesium, and macro- and micro-elements if needed; C) wait one growing season for minimal nitrogen applications then maintain minimal levels for 3-5 years; D) if in doubt about the structural integrity or survival of a tree remove it immediately; and, E) watch closely for pests and changes in tree structure. Preventative treatments may be advisable.

Conclusions

The quality of life of the tree and the tree owner is dependent upon the design and development processes being tree-literate and using biology-first design concepts. If stupidity is its own reward, in the past we have been well rewarded with poor quality trees on many development sites. Tree quality can be preserved, maintained and restored if we give trees a biological and ecological chance.

Literature Cited

1. Coder, Kim D. 1988. Maintaining tree quality. Grounds Maintenance 23(5): 22,63-64.
2. Coder, Kim D. 1989. Outline for selecting wooded lots and preventing tree damage during construction. University of Georgia Cooperative Extension Service miscellaneous publication 322. Pp.2.
3. Coder, Kim D. 1992. Plantings: The foundation of healthy tree life. Pp. 160-163 **In** Proceedings of the Fifth National Urban Forest Conference. American Forests, Washington D.C.
4. Coder, Kim D. 1992. Storm-damaged trees. Grounds Maintenance 27(2): 33,36,38,42.
5. Coder, Kim D. 1995. Total quality management of trees (TTQM). University of Georgia Cooperative Extension Service publication FOR95-2. Pp. 2,
6. Coder, Kim D. and W.H. Slack. 1990. Selecting wooded home sites. University of Georgia Cooperative Extension Service Bulletin 1029. Pp.15.
7. Craul, Phillip J. 1992. Chapter 7: Soil compaction and its amelioration. Pp. 210-243 **In** Urban Soil in Landscape Design. John Wiley & Sons, New York.
8. Gobster, P.H. 1994. The urban savanna — reuniting ecological preference and function. Restoration and Management Notes 12(1): 64-71.
9. Morris, L.A. and R.F. Lowery. 1988. Influence of site preparation on soil conditions affecting stand establishment and tree growth. Southern Journal of Applied Forestry 12(3): 170-178.
10. Perry, Thomas O. 1990. Conditions for plant growth. Pages 103-110 **In** Proceedings of the Fourth Urban Forestry Conference, P.D. Rodbell, editor. American Forestry Association, Washington, DC.

Effective Tree Retention in New Developments

Julian A. Dunster

Throughout the Lower Mainland of British Columbia, the public and political pressure to retain trees in new developments is high, but many well intentioned designs fail. Typically, tree retention efforts have saved second growth coniferous forests without sufficient concern for tree behaviour in high winds and rain; a common combination in this area. Subsequent tree failures and property damage suggest the need for rethinking this technique. Using a new approach, the taller second growth trees are carefully removed, minimising site disturbances and maximising retention of the important ecological attributes of the landbase. The selectively logged site is then fenced off with chainlink fencing and the area replanted. In this manner, the landbase is retained in relatively undamaged condition, along with the existing understorey plants and any advance regeneration. By planting replacement stock at the start of development, new trees become established early on. At the time of completion, often several years later, a new and healthy forest is already growing into the different site conditions, and important ecological attributes are already recovering. The approach is working well. It provides a safer solution while at the same time, retaining useful ecological and aesthetic features in the developed landscape at a fraction of the cost associated with more typical approaches.

The Lower Mainland of British Columbia encompasses a large valley bottom and floodplain area, bounded on the north shore by the Coast Mountain Range, by the Gulf of Georgia in the west, the Fraser Valley to the east, and leads down towards the international border with Washington State, and the Cascade Mountain Range in the south. The climate is maritime, with high annual rainfalls and seasonal wind storms coming down from the arctic, as outflows down the Fraser Valley, or as off–shore oceanic winds coming up the Oregon and Washington coastline.

Most of the remaining forest cover comprises second growth coniferous forest, typically less than one hundred years old, and dominated by Douglas fir (*Pseudotsuga menziesii*), western hemlock (*Tsuga heterophylla*) and western red cedar (*Thuja plicata*), with smaller components of red alder (*Alnus rubra*), bigleaf maple (*Acer macrophyllum*), vine maple (*Acer circinatum*), and occasional stems of paper birch (*Betula papyrifera*), black cottonwood (*Populus trichocarpa*), and bitter cherry (*Prunus emarginata*). Most of these second growth stands have developed on sites that were

Dr. Dunster is president of Dunster & Associates, Environmental Consultants Ltd., P.O. Box 109, Bowen Island, British Columbia, Canada V0N 1G0

logged and then burned around the turn of the century, and archival photographs show vast tracts of the Lower Mainland as smouldering remnants of a once grander forest. Many of the forests in the valley bottom have often regenerated after long periods of clearing and agriculture.

In recent years, public and political awareness about environmental issues in general, and of forests in particular, has developed rapidly. There has been a widespread and often vitriolic province–wide debate about the fate of our commercial forest land base. A spinoff effect of that debate has been greater focus on private land logging, especially when new subdivisions are being created. One of the more visible results of these debates has been a strong political response to retain trees in new developments (4). Sadly, much of this well–intentioned legislation has lead to poorly implemented tree retention efforts. The public and political prerogative to save trees has often been implemented at almost any cost. In many cases, developers, municipal staff, and members of the design professions are attempting to save almost anything that lives, in an attempt to be politically correct, but such an approach is already creating an expensive backlash. And, as is typical in so many other places as well, many of the trees retained suffer a wide range of abuses during development, thus further limiting the chances of successful retention.

Many of the trees retained are completely unsuitable for retention. Some are seen as small isolated pockets of forest that was once more extensive; they may look good on the drawing board – as pleasing patches of green ambience – but on the ground they look like a disaster. Worse still, are some of the individual trees, once part of a functional forest ecosystem, but in isolation, completely unadapted to conditions outside their original closed canopy forest structure.

The result of such poorly thought out retention efforts, is a legacy of failed designs, lost opportunities, and an ecologically impoverished landscape (3,5,6). Considerable amounts of money are being spent to retain inappropriate trees; buildings are moved or redesigned, density is lessened or traded, and designs are promoted with little understanding that the chances of success are small in the short term and virtually non–existent in the long term. In total, significant opportunities are being lost to integrate effective tree retention and ecologically valuable attributes into the new urban design. The knowledge needed for alternative approaches already exists (3,5,6,8), and includes considerable effort locally to treat the aquatic components of the landscape with far more care (1,2), although this too creates some interesting tree retention results.

Of equal concern, these ill–fated tree retention efforts cost additional money later on, usually after the development is completed and the performance bonds paid out. The new owners often face very expensive large tree removal costs, houses are being damaged as the trees blow over in the seasonal storms, insurance costs are rising, homeowners are losing their no–claims deductions, and the promise of long–term forested ambience does not materialise as intended.

A large part of these failed designs stems from a widespread lack of ecological understanding among the design professions. The "flat earth" planning approach (3) in which the land is first cleared of all obstacles, then developed, and then re–landscaped, is still a dominant paradigm in many cases. While many developers and design professionals often find this simpler to manage, it is an approach that greatly diminishes the value of the finished development, and often decimates many important ecological values.

More sophisticated approaches recognise the value of retaining ecological features linked throughout the landscape (5,7). Some (sadly, very few) attempts are being made to incorporate these into new developments. But again, many efforts fail due to lack of ecological understanding among other members of the design team, such as planners and engineers, the architects and landscape architects, and the politicians. As a result, it is difficult to achieve the full ecological potential of many sites, although there is little doubt that this could be achieved with greater understanding about tree biology and ecology in general (3,6,8).

The *status quo* approach to retaining trees within developments has largely ignored the biological realities facing us. Many of the second growth forests are composed of uniform, even aged coniferous stands; widespread areas of codominant trees, all reaching for the light; and typically tall, spindly stems having most of their foliage in the upper 15 to 20% of the trunk. The overlapping canopies support each other at this stage of stand development, and consequently, the undisturbed forest has some windfirmness and structural integrity.

Mixed in with these stands are the occasional veteran from the previous forests, usually very dominant Douglas firs that have much more pronounced open-grown forms, denser branching patterns in all directions, dominance over the surrounding forest, larger trunks, and better root systems that have developed over longer periods of time in response to many years of varying weather conditions. Other variations include large open-grown or codominant bigleaf maples or alders, and the occasional open-grown birch. More typically, the alder, birch, and cherry trees are pioneers and are tall and spindly, with many of them in biological decline (although still providing important ecological attributes to the site as a whole). Mixed in with the overstorey, are examples of most species in a suppressed form; the conifers typically at 50 to 60% of the overstorey height, and the deciduous trees growing in small openings throughout the stands.

Because most designs must be placed within the confines of this fairly extensive second growth forest, new forest edges are always being created. Unfortunately, the new edge trees, and especially the conifers, formerly surrounded by codominant cohorts, are inherently unstable. The problem is greatly exacerbated when the entire tree retention area is composed of new edge trees. Experience has clearly shown that these new edge trees are extremely difficult to retain. The tall conifers, and especially the hemlock, blow down or snap off in the seasonally high winds. The alders snap off or scorch in the sun. The bigleaf maples shed branches, and in total, much of the planned tree retention ends up as firewood and mulch. In recent years there have been an increasing number of incidences where property has been damaged, and although personal injuries have not been severe so far, many of the current designs are unknowingly creating *de facto* hazard trees.

Retaining this second growth forest with any degree of long–term success requires that much larger areas of forest be set aside from development, and that any new developments be set well away from the forest edges. The 90 to 100 year old conifers in this area can grow 55 to 60 meters high, although most are in the 45 to 50 meter height class. This means that any such set back zone, provided to ensure a minimum of safety from falling trees, needs to be at least one tree height wide if semi–mature conifers are present.

However, the high price of land in the Lower Mainland effectively precludes this approach, because it would remove considerable amounts of developable land. As a result, many developers and municipalities accept the risks involved, perhaps, unknowingly. Theoretically, the ideal shape for retention would be circular to minimise the amount of perimeter edge, with local adaptations of shape to try to accommodate the prevailing wind patterns. However, most of the areas being used are still far too small to be successful, regardless of their shape, and unstable edge effects lead to extensive windthrow, and expensive clean up costs and insurance claims.

An Alternative Approach

In a break from these more normal approaches, a different system is being used with considerable success. This alternative approach is predicated on the following basic needs:

1) to retain an ecologically functional land base that is capable of growing trees well into the future, and that can be integrated into the overall landscape as a means of retaining some useful ecologically viable features and linkages.
2) to eliminate the hazard trees on the site and thus reduce long–term management, liability, and insurance costs.
3) to provide a site upon which a healthy new forest can be created with less expenditure of capital, and less short and long–term management costs.
4) to recognise the importance of entire ecosystems and not just the big tall trees that create a big impression (especially when they fall down).

The first site where this approach was tried was very small, and covered with conifers in the 45 to 50 meter size class. These tall trees were retained as long as possible on the site to retain some visual effects, but once the building work came within striking distance, the trees were removed. The tree removal stage is probably the most crucial link in the entire approach. The specifications developed are very strict and have to be rigorously enforced.

First of all, the land base to be retained (not the tall trees) is defined in the plan and then surveyed and clearly delineated on the ground. A boundary fence or some other form of marking (a continuous polypropylene rope works well) is erected. Site clearing, grubbing and soil stripping can take place outside of this boundary, but everything within it is left untouched.

At a convenient stage in the site clearing, the tall trees are assessed and marked by the arborist. They are then carefully cut down, aiming wherever possible to have them fall out of the tree retention area so that the understorey vegetation suffers minimal disturbance. In some cases, a high line is tied to the trees by a tree climber and the tree is directionally felled to avoid damage. In other cases, excavators have reached in and pushed trees out of the retention area. The important criterion is that all tree felling is done so as to create minimum ground disturbance.

Once the trees are down, they are extracted. Machinery is permitted to sit at or just beyond the boundary and reach in, but under no circumstances is any machinery permitted to enter on to the land base to be retained. All trees are required to be extracted in a suspended lift; that is, they must be clear of the ground and not skidded across it. Realistically, this is not always feasible, especially in the larger areas. In such cases, a snag tree is created at the edge and a block hung high up so that a wire rope can be used to gain partial lift of the logs as they are skidded out. Once at the edge they are then lifted up and out. Some operators have also added a temporary grapple claw on top of the excavator bucket to facilitate extraction.

The trees are cleaned up outside the boundary, and the tops and branchwood chipped on site, with the resulting mulch then being broadcast back into the remaining area, and especially along the edges where soil disturbances are highest. Some hand cleaning work may be necessary after extraction to clean up minor debris, but generally the area is left alone.

At this stage, the land base is securely fenced off, using a chain link fence placed along all accessible boundaries. In the following fall or early spring, replanting takes place, typically using 1.5 meter coniferous stock, or similar sized deciduous materials or whips. Use of smaller material seems to be very prone to browse by deer, while larger stock goes into a longer period of transplant shock, costs considerably more, and ultimately, is often outgrown by the smaller stock.

By the end of this process, there is a clearly defined, and largely undisturbed landbase upon which a new forest can be created. The existing understorey vegetation is retained, along with any advance regeneration that already existed on the site. The newly planted stock has much more favourable microclimatic conditions for establishment. i.e. there is good soil that is not compacted and thus retains some of the existing moisture regimes; there is plenty of shade and vegetation to prevent desiccation; and, from an ecological point of view, there is considerable structural diversity at the ground level, providing many important niches for wildlife and the development of new vegetation.

To date the new approach has been implemented on several sites, now ranging from a few hundred square meters to many hectares. There are many significant advantages including:

- the hazard trees are removed before targets are established;
- the areas to be retained for future forest can be much more effectively incorporated into designs because the designer is less constrained by the "protect the big trees" syndrome;
- many more useful ecological functions can be retained across a larger landscape, and landscape linkages and stepping stones of habitat can be incorporated to meet a wider array of users (birds and animals) or uses.
- the costs are considerably less, since the developer gains the value of the logs extracted at the clearing stage, can gain the design and density wanted, and has no liability. The long–term management costs are also much less, because the retained land base and replanted trees are designed to be left alone with no maintenance or other interventions, as a long–term green space. And, the capital costs of landscaping are considerably reduced, since the main requirement is for replanting stock of a smaller size class, rather than more expensive (larger) stock and its associated higher installation costs.
- a final advantage is that the approach creates healthy new forests, with some valuable ecological integrity, that can adapt and develop into the new site conditions. In the coming decades, these areas will provide a much more lasting forested ambience than those created under the more traditional approach.

Critics might argue that the visual appeal of these new forest areas is much less, because the tall trees have been removed, and the replanted areas may not develop to their full potential. The response is to note that the visual appeal of new edges in second growth forest is very low. The close up viewer is confronted with a sea of hazardous edge trees that look like poles, little if any tree foliage, and in some areas, little or nor understorey vegetation. By contrast, the alternative approach provides a developing healthy young forest at eye level, and one has only to watch the annual growth rates of our native species to see that a meter or more of growth is quite typical once the trees are established. If the specifications are closely followed, the undisturbed landbase has far more visual appeal as a small natural area than the more traditional – ball and burlapped stock surrounded by bark mulch – approach to landscaping.

Overall, this alternative approach of retaining an undisturbed forest landbase, rather than attempting to save unsuitable trees, has considerable merit for many sites throughout the Lower Mainland. The approach is gaining acceptance as the results can be more clearly proven. And, although politicians and designers are often reluctant to see the taller conifers removed, there is a growing realisation that careful planning and implementation can yield a better solution. The key to success is an undisturbed land base, a much more thorough understanding of tree biology, and forest ecology, combined with a design team willing to adopt innovation.

Literature Cited

1. Anon. 1993/94. Stream Stewardship. A guide for Planners and Developers. Vancouver: Department of Fisheries and Oceans.
2. Chilibeck, B. 1992. Land Development Guidelines for the Protection of Aquatic Habitat. Vancouver: Department of Fisheries and Oceans.
3. Dorney, R.S. 1989. The Professional Practice of Environmental Management Ed. Lindsay Dorney. New York: Springer Verlag.
4. Dunster, J.A. 1994. New legislative ways of protecting trees in municipalities: An overview of the British Columbia approach. J. Arboric. 20(2):109–113.
5. Dunster, K., and Dunster, J. 1992. The Nature of Burnaby: An Environmentally Sensitive Areas Strategy. Dunster & Associates.
6. Hendler, B. 1977. Caring for the Land. Planning Advisory Service Report No. 328. Chicago: American Society of Planning Officials.
7. Hudson, W.E. 1991. (Ed). Landscape Linkages and Biodiversity. Washington, D.C.: Island Press.
8. Marsh, W.M. 1983. Landscape Planning. Environmental Applications. Massachusetts: Addison– Wesley.

Pre-development Tree Assessment

Jeremy Barrell

In Britain there is intense competition for space; people want to live in it, industrialists want to build factories in it, farmers want to farm it and trees want to grow in it. On a small island, with an increasing population, there is quite simply not enough space to go round, so there is a need to plan for who has what and what goes where. In response to these pressures, Britain has evolved a sophisticated array of planning laws and regulations. In almost every aspect of our daily life, the British have to deal with some form of control over the whole range of our activities, from how and where we construct our buildings to how many cows we can have running around in our fields!

With an ever increasing awareness of the environment, our urban trees are also subject to a high level of control. This is primarily achieved through *The Town & Country Planning Act* which allows for the protection of trees and woodlands on grounds of amenity by law. It also places a statutory duty on municipalities to actively consider existing trees and future landscaping when determining applications for permission to carry out new development. If they are not satisfied that trees have been adequately considered in a construction site application, then they can refuse permission. Refusals of permission cost developers money and stop them from making more money. This is the engine that has driven the evolution of tree management methodologies on development sites to such an advanced level in Britain.

In order to meet the requirements of the legislation, advisory guidelines have also evolved to a high level. One of the most frequently used and highly respected British publications dealing with the management of trees on construction sites is *British Standard (BS) 5837 'Guide for Trees in Relation to Construction' - 1991* [2]. Despite its high status, the British Standard is more of a guidance reference than an absolute set of rules. It represents the starting point from which individual prescriptions can be evaluated by the tree expert (arboriculturist). As with all tree matters there are no simple recipes and the arboriculturist is by far the most important component of the tree management equation. The British Standard offers a framework for managing trees throughout the development process covering pre-development tree assessments, protection during construction, appropriate tree/building separations, changes around existing trees, landscaping and final project hand over. Whilst BS 5837 is not based on any specific research, it is relevant that more recent work, most notably by Matheck [3], supports the BS guidance, further enhancing its credibility.

Mr. Barrell is the owner of Jeremy Barrell Treecare, 25 Hightown Gardens, Ringwood, Hampshire, BH24 3EG, England.

Pre-development Tree Assessments

Pre-development tree assessments are a means of establishing the relative usefulness of existing trees on potential construction sites. They involve allocating trees to categories of importance to help identify the most sensitive areas of the site. They can range from a simple assessment, based on the visual importance of each tree, to more complicated evaluations based on life expectancy and usefulness. With this information, reasoned judgements can be made on which layout design will optimise the retention of the best trees. This is a vital pre-requisite to establishing what precautions are necessary to ensure their successful retention.

In the British planning system a high importance has been placed on the value of trees within new developments. This evolution to date has been difficult because the planners designing these new developments are not tree experts and the arboriculturists presenting tree information are not planning experts. The challenge has been to evolve a methodology that allows the transfer of information from the arboriculturists to the layout designers without too much distortion. Trees can only be effectively considered in the overall scheme of a new construction site when the layout designers are presented with easy to interpret information. In practice, the most effective presentations have proved to be simple with a strong visual content.

The planning process begins when the decision to develop is made and finishes when the completed development is occupied. If trees are to be properly incorporated into the development, there should be at least six separate stages of arboricultural input between the beginning and the end. Pre-development tree assessments are only one small part in this overall process (Figure 1) but are the most important because they provide the basis on which all the subsequent decisions are made.

The simplest categorisation of trees are based on objective criteria such as size, species or position within the landscape. These assessments can be carried out with a reasonable degree of consistency by anyone with an average level of tree knowledge. However, the more sophisticated methods involve subjective criteria such as life expectancy, hazard potential and usefulness. Their proper use demands extensive practical experience with trees and a detailed technical knowledge of tree biology. These more sophisticated methods should only be carried out by professionals experienced and qualified in arboriculture who have been trained in their use.

As pre-development tree assessment methodologies have evolved, a number of problems have materialised:

Amenity. The amenity value of trees is probably their most obvious asset but also one of the most difficult to incorporate into an effective pre-development assessment. The problem arises from the fact that the largest trees with the highest amenity value usually tend to be the oldest with the shortest life expectancy. In most situations it is inappropriate that the long

Figure 1

Where pre-development tree assessments fit into the planning process

Stage in Development	Arboricultural Advice

Decision to Develop

↓

Stage in Development	Arboricultural Advice
Land survey	
→ Tree survey	Pre-development tree assessment ←
Preliminary layout design	
Fine tuning	Advise on minor adjustments to layout to minimise effects on retained trees
Finalisation of preferred layout	• Prepare schedule of tree works • Advise on positions of protective fencing

↓

Submission to Municipal Authority

↓

Stage in Development	Arboricultural Advice
Negotiations and possible modifications to comply with municipal authority requirements	Advise on impact of further layout changes

↓

Consent to Develop

↓

Stage in Development	Arboricultural Advice
Construction of development	Site control: • supervision of tree works • supervision of protective fencing
Completion of construction	Signing off tree responsibilities

↓

Occupation of Finished Development

term layout design should be significantly influenced by trees which will only be present for perhaps a few more years at the most.

Small trees. With present day abilities to easily move small trees or replace them with virtually identical semi-matures, it is inappropriate that they should dictate the long-term layout of a new construction site. Small or young trees must be specifically dealt with if a pre-development tree assessment methodology is to be effective.

Subjectivity. A major drawback of methodologies based on subjective criteria is the difficulty in consistently arriving at the same answer with different assessors. This problem can never be fully addressed but if a methodology is to survive, it must go some way to ensuring that more or less the same answer can be consistently achieved.

Presentation of Information

The way information is presented is very important. Effective presentation will facilitate understanding and interpretation. Poor presentation will foster confusion and misinterpretation. Pre-development tree assessments are about the simplification and transfer of complex tree information. The objective is to carry out these processes in a way that minimises any change of content of that information. Tree experts simplify the information and transfer it to non-tree experts who use it. It is the responsibility of the arboriculturist to make sure that the information is easily understood by the end users. There are two important elements of information presentation; the tree schedule and the site plan.

Tree schedule. All pre-development survey information must be presented as a tabular tree schedule. Tables allow an overview at a glance and enable specific items of data to be extracted almost instantaneously without wading through pages of text. Figure 2 is a sample tree schedule illustrating the range of data that must be incorporated in a full pre-development survey. The key features are:

- Height, spread and trunk diameter all help indicate tree size in a complete way
- Trunk diameter, vigour and maturity are all needed to reference BS 5837 to establish the positioning of protective fencing
- Assessment categories establish the relative importance of each tree
- The 'Comments' column provides the space to elaborate; perhaps the further clarification of the reasoning behind the assessment category allocation or any special characteristics related to amenity
- Explanatory notes after the schedule table should always be included to provide further information on botanical names, abbreviations and background to the assessment methodology.

Site plan. The site plan does not provide any new tree data in addition to what is already included in the tree schedule. Pictures allow the overview to be presented instantaneously; colour makes the data more interesting and increases the amount of information that can be presented on one plan.

- Assessment categories in colour for easy and rapid reference
- Crown spreads drawn as they are on the ground and not the usual perfect circles
- Include trees outside the site but close to the boundaries, especially where their crowns and roots extend into the site
- Height and diameter details next to each tree is useful to provide a quick guide to their size
- Existing features and levels information are vital to enable complete incorporation of the tree data into the layout design.

Please note the plan in Figure 2 is shown at a scale of 1:500 to provide an over view. In practice, scales of 1:200 are generally more appropriate.

Presentation is the key to the effective transfer of information. If it is misinterpreted then it is the arboriculturist's failure. Arboriculturists should keep the following points in mind to help improve the effectiveness of this transfer: 1) They are trying to convert complex tree information into a simple form and transfer that information to non-tree experts with clarity. There should be no jargon. It should be easy to interpret and it should be interestingly presented. 2) Layout designers need full information to make the best use of the site. The common reality is that they will be preparing the layout without having visited the site. The quality of tree data directly affects the quality of the development. Arboriculturists have an important role to

Figure 2

Example Tree Schedule for a Proposed Construction Site

Tree No	Species	Height (m)	Crown Radius (m)				Diameter (cm)	Maturity	Vigour	Assessment Category	Comments
			North	South	East	West					
18	Oak	20	5	3	3	3	90	Mature	N	3d	Significant storm damage in upper crown but retainable in short term providing valuable boundary screening
19	Beech	20	12	10	3	7	95	Over mature	L	4c	Severe die-back in upper crown and extensive advanced decay in lower trunk
20	Beech	14	14	10	6	2	24*	Young	N	1a	Slightly suppressed by T19 but well formed with good future potential

Explanatory Notes

- **Botanic Tree Names:** Beech : *Fagus sylvatica*; Oak : *Quercus robur*

- **Abbreviations:** m : Metre; cm : centimetre

- **Height** - tree height is normally visually estimated and recorded in metres. If appropriate, it can be measured optically and these measurements are indicated with a '*' after the number.

- **Crown Radius** - the average crown radius from the centre of the trunk to the tips of the live lateral branches is estimated and recorded in metres.

- **Diameter** - trunk diameter at 1.3 m above ground level is measured with a diameter tape and recorded in centimetres. Where trees have multiple stems this will be recorded as an 'M' or the diameter of each stem will be listed separately, whichever is most appropriate.

- **Maturity** - the maturity was broadly assessed as over mature (beyond normal life expectancy or in an age related state of decline), mature (last one third of life expectancy), maturing (one third to two thirds life expectancy) and young (less than one third life expectancy) to comply with the definitions in Table 1 of BS 5837.

- **Vigour** - an indication of the health of the tree for use with Table 1 of BS 5837. N = normal vigour and L = low vigour as used in that table.

- **Assessment Category** - the contents of this column above and the explanation here depend on the type of assessment being carried out. In all cases a description of all the categories used must be included at this point. Some further explanation of the methodology employed as background information may also be useful to the end users of the above data.

play. 3) There will be other end users of this information. Construction sites are often the subject of detailed arguments and it is likely that a non-tree expert may have to make decisions relating to trees. Information should be easily digestible by non-tree experts. The quality of their decisions will be directly influenced by the quality of the tree information that is placed before them.

SULE is the State of the Art

SULE is an acronym for **S**afe **U**seful **L**ife **E**xpectancy. Provisional details were first published two years ago [1] and its evolved form now represents the current state of the art in pre-development tree assessment methodology for construction sites in Britain. Its central theme is that in a planning context the length of time a tree can be expected to be usefully retained is by far the most important long-term consideration. SULE provides a structured methodology for systematically assessing the key aspects of retaining trees near people.

SULE prioritises individual trees within a defined area and enables value judgements to be made to determine which are the most suitable for retention when there is a shortage of space. The methodology describes the detail of transferring tree information from the arboriculturist collecting it on the ground to the layout designer using it in the office. SULE is the vehicle for this information transfer and its strength is that it minimises the opportunity for misunderstandings between the collector and the user. The SULE methodology provides two levels of guidance; detailed explanations for the arboriculturist and a general overview for the end users to enhance their background understanding. The end result is an easy to understand categorisation of the trees that enables the layout designer to make the best use of the site.

Each tree on the development site is inspected by the arboricultural surveyor and individually assessed for its SULE. This is recorded on a tabular tree schedule along with other appropriate information and visually as a coloured site plan. It is then presented to the layout designer, who can identify the most and least important trees on the site and design the layout around them. This allows the full impact in terms of tree loss to be systematically evaluated for different layouts in a reasoned way. It leads to the selection of a preferred layout optimising the use of the existing trees.

SULE is about tree life expectancy, and how long they can be expected to be retained safely and usefully. It is based on a number of obvious management assumptions, and the fundamental principles of safeness and usefulness.

To make effective management decisions it is necessary to have a core group of basic objectives. It is assumed that the following basic principles apply to most urban tree situations: 1) Management is preferable to no management; good management is preferable to bad, 2) Safety is the absolute priority, 3) Reasonable management costs is an important secondary objective, and 4) Sustaining amenity is an equally important secondary objective.

Good management implies that only accepted tree care practices will be used, and that any operation that is known to adversely affect long term tree health would generally not be appropriate. For example, significant reductions of mature trees would not be generally considered as good practice or acceptable. It is also not good practice to artificially keep trees in a clearly unsuitable position. For example, in most cases it is artificial and contrary to good management to keep large trees in small spaces because of the need for continual pruning. Good management will strive to achieve the right tree for the site and seek to avoid practices that adversely affect tree health.

In all situations close to people or property, safety has to be the priority consideration above economics or amenity. The measure for action is hazard potential. Hazard potential is related to tree size, tree structure and the number of targets that could hit [4]. As trees grow bigger their potential to cause damage increases it. As tree structure becomes more suspect, so the chance of failure increases. As the number and value of targets that could be hit increases, so the potential cost of damage or injury increases. The priority when managing trees with a high hazard potential should be to reduce the risk to an acceptable level. This can be achieved through removing the tree, removing the targets or treating the tree.

Trees are useful in an urban environment if they are cheap to keep and they contribute to amenity. They become less useful as maintenance costs become excessive and they begin to have a negative effect on amenity by interfering with better trees or inhibiting the establishment of new trees.

Economics of management. Urban tree management will always have financial constraints with limits on how much expenditure is acceptable. It is not reasonable nor useful to manage trees in a way that costs more than other suitable options. Retaining trees at an excessive management cost is not normally acceptable unless there are very special reasons such as rarity or some commemorative value.

Sustaining amenity. In most urban situations the whole point of having trees is to make a significant contribution to the amenity of the local environment. A major part of this amenity is through their size and impact on the landscape. The larger they are, the greater tends to be their importance. It follows that good management should seek to maximise this contribution and minimise the impact of necessary maintenance. It is a common feature of localised plantings or groups that all the trees are of a similar size or age. The implications of this are that many trees will reach maturity and need removing at about the same time, resulting in rapid changes to the local landscape. It is inevitable that as trees mature they will need to be removed and replaced. Good management should seek to spread these operations over the whole rotation, thus reducing the number and impact of removals at any one time. Sustained amenity is achieved by establishing a range of age classes within a local population, from new planting right through to mature trees. In most situations this can be achieved by removing and replacing trees that are not performing because they are not suited to the site or they are interfering with better trees. Trees are useful until their retention compromises this principle of sustaining amenity.

It is not possible for just anyone to go out and undertake subjective tree assessments competently. Extensive experience in day to day practical tree management and a thorough grasp of theoretical arboriculture are basic requirements. In addition to this, it is then necessary to undertake further training in order to become competent with a particular method. SULE is not easy to understand or undertake, and it is certainly not possible to just go out and do it. There are numerous grey areas and apparent ambiguities that need to be encountered and understood before SULE assessments can be carried out with confidence and consistency. When familiar with the method, assessors should be able to quickly go through a structured sequence of considerations mentally to arrive at a SULE figure for each tree. For those who are unfamiliar with the method it is necessary to systematically learn how to do this. Structured training under the supervision of an experienced SULE instructor is the quickest and easiest way to become competent in its use.

Summary of the SULE Assessment

In order to understand the whole, it is necessary to identify the individual component stages in the methodology. Each stage should be considered in a systematic way before a final SULE figure can be arrived at for each tree. SULE is based on tree life expectancy, safety and usefulness. For every assessment these need to be considered separately in the order listed below:

1. Estimate the age of the tree
2. Establish the average life span of the species
3. Establish if that average life span needs to be modified because of local environmental circumstances
4. Estimate life expectancy (life expectancy = average modified life span of species - age of tree)
5. Consider how health will affect safety
6. Consider how tree structure and size will affect safety
7. Consider how location will affect safety
8. Establish safe life expectancy (safe life expectancy = life expectancy modified by health, structure and location)
9. Consider economics of management - costs must be reasonable
10. Consider adverse effects on better trees
11. Consider sustaining amenity - making space for new trees
12. Establish SULE (SULE = safe life expectancy modified by economics, effects on better trees and sustaining amenity).

At first sight this may seem long winded and complicated, especially to those who are familiar with carrying out tree assessments. However, for novice SULE users this is a vital checklist to make sure that no steps are left out. As you become familiar with the method and how it all fits together, dealing with the individual steps becomes almost second nature and an assessment can be carried out very rapidly indeed.

Once the SULE in years has been assessed, it is a simple matter to place the tree into the appropriate SULE category and record it on the tree schedule. Each SULE category has a number of sub-divisions which help to clarify the reasoning behind that particular assessment. It is important to record the relevant subdivision to aid interpretation of the information. Suggested categorisations and subdivisions are included below:

Long SULE. Trees that appeared to be retainable at the time of assessment for more than 40 years with an acceptable level of risk. A) Structurally sound trees located in positions that can accommodate future growth. B) Trees that could be made suitable for retention in the long term by remedial tree care. C) Trees of special significance for historical, commemorative or rarity reasons that would warrant extraordinary efforts to secure their long term retention.

Medium SULE. Trees that appeared to be retainable at the time of assessment for 15 to 40 years with an acceptable level of risk. A) Trees that may only live between 15 and 40 more years. B) Trees that may live for more than 40 years but would be removed for safety or nuisance reasons. C) Trees that may live for more than 40 years but should be removed to prevent interference with more suitable individuals or to provide space for new planting. D) Trees that could be made suitable for retention in the medium term by remedial tree care.

Short SULE. Trees that appeared to be retainable at the time of assessment for 5 - 15 years with an acceptable level of risk. A) Trees that may only live between 5 and 15 more years. B) Trees that may live for more than 15 years but would be removed for safety or nuisance reasons. C) Trees that may live for more than 15 years but should be removed to prevent interference with more suitable individuals or to provide space for new planting. D) Trees that require substantial remedial tree care and are only suitable for retention in the short term.

Removals. Trees that should be removed within the next 5 years. A) Dead, dying, suppressed or declining trees through disease or inhospitable conditions. B) Dangerous trees through instability or recent loss of adjacent trees. C) Dangerous trees through structural defects including cavities, decay, included bark, wounds or poor form. D) Damaged trees that are clearly not safe to retain. E) Trees that may live for more than 5 years but should be removed to prevent interference with more suitable individuals or to provide space for new planting. F) Trees that are damaging or may cause damage to existing structures within 5 years. G) Trees that will become dangerous after removal of other trees for the reasons given in A) to F).

Moved or replaced. Trees which can be reliably moved or replaced. A) Small trees less than 5 meters (m) in height. B) Young trees less than 15 years old but over 5 m in height. C) Trees that have been regularly pruned to artificially control growth.

There will always be cases that do not fit neatly into one category. It is quite acceptable to indicate this in the 'Category' column of the tree schedule and record the problem with allocation in the 'Comments' column. Remember, the objective is to provide reliable information and that means accepting that no methodology is going to be able to realistically cover every possible situation. Sometimes, long hand descriptions are the only way!

The Future

In the USA the pressure is now on. It is no longer acceptable to just bulldoze trees in the name of progress. Increasingly, when developing new countryside lots and redeveloping established inner city sites, existing trees will have to be given due consideration. In Britain we have been under these pressures for many years and have evolved the expertise and systems to ensure that trees are given the level of care that is demanded. The willingness of your public and private sectors to take advantage of this expertise will impact directly on your ability to cope efficiently with these rapidly materialising requirements in the USA.

SULE is the cutting edge of tree assessment methodology. It is now better documented and more effective in use than any other method currently available. SULE is an absolute, basic pre-requisite in all construction site situations if trees are to be given due consideration. There are no reasons why it cannot be applied to the circumstances that are developing in the USA at the moment. This presentation has been designed to introduce the basic concepts of the methodology; to tell you that SULE is here. It is tried and tested in practice and we have the expertise to get it up and running in the USA.

Since SULE was originally conceived 12 years ago it has been continually updated through extensive testing in the field and use in the British planning system. A series of workshops throughout Britain during the last two years has highlighted the areas that require modification. Full and detailed explanations of the SULE methodology based on this feedback will be unveiled in a paper to be presented at the main ISA Conference at Hilton Head later this year. However, experience in Britain has shown that information alone will not be sufficient to ensure its competent use. Training will be necessary and it is intended to run a series of workshops in the USA in 1996. These will be based on an easy-to-absorb mix of field practicals and classroom sessions, backed up by a delegate pack with full written explanations and a regular update service.

Literature Cited

1. Barrell, J.D. 1993. Pre-planning tree surveys: Safe Useful Life Expectancy (SULE) is the Natural Progression. Arboricultural J. 17: 33-46.
2. British Standards Institution. 1991. Guide for trees in relation to construction. BS 5837: 1991. BSI, Linford Wood, Milton Keynes, MK14 6LE.
3. Matheck, C. & H. Breloer. 1994. Field guide for Visual Tree Assessment (VTA), Arboricultural J. 18: 1-23.
4. Matheny, N.P. & Clark, J.R. 1993. A Photographic Guide to the Evaluation of Hazard Trees in Urban Areas, 2nd Edition, International Society of Arboriculture, PO Box GG, Savoy IL 61874.

Trees and Development Conflicts: The Importance of Advance Planning and Site Control in Tree Preservation Plans

Dealga O'Callaghan and Michael Lawson

The conflict between the needs of developers to maximise returns of expensive building land and the environmental pressures for the preservation of trees in a healthy condition can be satisfactorily achieved without compromising arboricultural integrity. A process of site assessment combined with development control provides both planners and developments with effective solutions to the traditional conflicts. The process fits well into the existing planning framework in Britain and meets all the recommendations of the relevant British Standards and Government advice while remaining familiar and user friendly to the developers. It is a unique but simple approach from which all parties gain and trees are retained in good health.

Protection and preservation of trees on construction sites is a perennial problem for arborists, planners and developers. The conflict between the need to maximise scarce building land and the environmental pressure to retain as many trees as possible often sets the construction industry at odds with the requirements of planning Authorities and the needs of trees.

This conflict has been recognised in Britain for many years where the planning process has evolved to encompass and preserve trees from its foundation in present form in 1947 (5). The evolution of the planning system has resulted in the establishment of a need for Development and Structure Plans which set the planning priorities for local authorities (Cities, Boroughs, District and County Councils) (5). These have now largely been replaced by the Unitary Development Plans or UDPs.

Trees in Britain can be protected legally under the Planning Act by the creation of Tree Preservation Orders (TPOs) which have the effect of preventing any work, including the cutting down, up rooting, lopping, topping, or willful destruction of a tree without prior written planning consent, which can only be obtained from the Local Planning Authority (LPA) (7,10) which is effectively the Municipal Planning Department. In addition trees on sites which are to be developed can be protected by means of conditions of planning in addition to or in place of TPOs (9). Such conditions of planning are enforced and violation can lead to stop orders, legal proceedings and substantial fines (7,8,10).

The scarcity of development land within the United Kingdom and the

O'Callaghan Associates Ltd., Valleyfield House, 1a Stratford Road, Aigburth, Liverpool, L19 3RE, England.

policies of recent government administrations to hold a general presumption against development in the countryside and greenbelts and in favour of the use of "infill" sites all combine to put the existing urban tree population under intense pressure from development (7). Recycled sites often have mature trees and the premium land prices lead developers to seek permission for as many units as possible in order to obtain a return on the very high cost of development land. The combination of these pressures and the requirements of the planning policy guidelines produced by the central government have resulted in a series of standards concerned with trees in relation to construction, the most recent of which is the British Standards Institute *Guide for Trees in Relation to Construction, BS 5837:1991* (2).

The Application of the British Standard BS 5837:1991

This standard evolved from an earlier *British Standard Code of Practice for Trees in Relation to Construction* which dated from 1980. The new standard is a major advance and encompasses many new developments in tree biology and tree care practices. Application of its recommendations improves the prospects for the healthy survival of trees close to new construction (3). In addition it gives guidance on suitable replacement species.

The rationale and methodology contained within the standard is advanced and requires a considerable degree of interpretation, particularly in relation to the minimum distances for protective fencing around trees recommended in Table 1 of the standard (2). These distances are calculated with reference to the age and relative vigour of the trees as well as to size as represented by the diameter (dbh) (2). It is in this area that planners are often at their most sceptical as they have difficulty believing that the distances are adequate to ensure the healthy survival of the trees despite advice from their own "in house" arboriculturists. However, recently Mattheck and Breloer (4) have shown that there is little if any failure of trees whose roots are damaged or cut at distances equal to or greater than those recommended by BS 5837:1991 although this is dependent to some degree upon the remaining root zone being protected and viable.

The British Standard also gives guidance on the type of pre-development tree survey which should be undertaken and how trees should be classified as those whose retention is highly desirable through moderately desirable to those which should be removed. The standard is a good guideline, but it is only that, and attempts have been made to refine and develop the pre-development survey categorisation to a system based on the safe useful life expectancy (SULE) of trees (1).

However, the problems for planners lie in the fact that too many arboriculturists are locked into the concept of the *Tree Survey* as a pre-development tool when what is obviously required is more than a survey listing trees to be removed and trees to be retained. The standard is a set of guidelines that need to be interpreted by the arboriculturist for both planner and developer. What the Standard actually requires is an Arboricultural Implication Study (AIS) which in addition to listing trees to go, etc., also discusses all the arboricultural implications of the development. The AIS provides the planners with all the detail they require to assess a development proposal correctly. Without it, the result is confusion, confrontation, and often, a refusal of planning consent.

It is emphasised that the Standard intends the user to implement a comprehensive assessment. However, the term "tree survey" has often been misinterpreted as meaning simply a list of trees.

The Planning Process in Britain

Before dealing with the form and role of the AIS, it is necessary to look in some detail at how the planning process works in the United Kingdom. The mechanisms are set out in Figure 1 and may at first glance appear intricate and cumbersome. However, the process is quite efficient and does work.

The Planning Acts. The over riding piece of legislation is the 1990 Town & Country Planning Act (9) which effectively governs all planning activity. It is supported by various sets of regulations and, because it is effectively a consolidation Act replacing the previous Acts of 1947 and 1971, it was not possible to make any substantial legislative changes when it was enacted (4). A separate Act of Parliament, the 1991 Planning and Compensation Act (8) provides for new legislation and particularly for changes in means of enforcement of the provisions of the 1990 Town & Country Planning Act and the penalty for violations.

The Planning Act is interpreted through various government advices, circulars, planning policy guidance notes (PPGs), etc., but in addition, Section VII of the Act deals specifically with trees and tree preservation. All of these place a duty upon the Local Planning Authorities to have policies on tree preservation, and tree and woodland management and to make provision for new plantings. These are contained within the adopted Local or Unitary Development Plans.

Within a Local Planning Authority there are basically two divisions, the Forward or Strategic Planning Group, who deal with the preparation and formulation of the Local Plans, UDPs, etc., and the Development Control Section, whose responsibility it is to assess all applications for permission to develop. Development control must assess all applications in relation to the existing adopted local plan and any emerging plans and their policies, government advice, etc., and must consult with an appropriate number of the various statutory bodies in relation to applications. These are listed in Figure 1 and include the Highways Department, Environmental Services, possibly the National Rivers Authority, the Council Arboricultural Officer, who may or may not be in the development control section, possibly with the Forestry Authority, if woodlands or forests are involved, with the Landscape Section

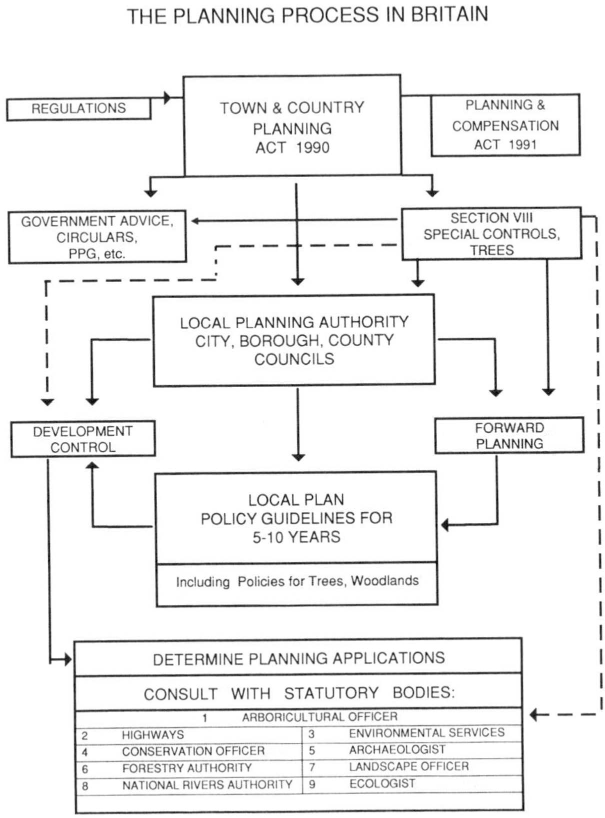

Figure 1. Chart illustrating how the planning process works in Britain in the 1990's. The shaded areas indicate the arboricultural aspects of the process.

and increasingly with local archaeologists.

Having consulted all of the above, the Case Officer reports to the Chief Planning Officer (CPO) and to the Planning Committee who ultimately vote on the decision having taken heed of their officers reports and advices.

Progress of an Application for Planning Permission. In order that development proposals can be assessed, an application has to be made for planning consent, to the LPA. The progress of the application depends upon many factors, not the least of which is the environmental sensitivity of the site. Often this hinges upon how many trees may be lost to accommodate the development proposal.

The ideal scenario is illustrated in Figure 2 where a developer really needs to put a design team together to formulate the application. Where there are trees on the site, this should always involve a consulting arboriculturist. The arboriculturist is the key player, he or she decides the extent of the site that can be developed (development window) in relation to

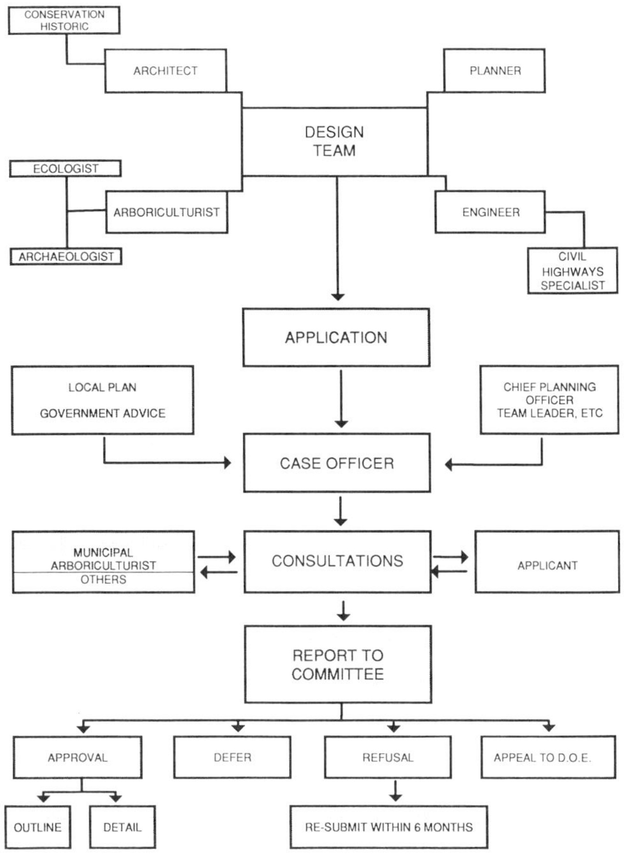

Figure 2. A chart illustrating the progress of an application through the planning system from initial design, through to approval or refusal. The shaded areas indicate the arboricultural aspects of the process.

what trees must be retained on site at all costs in line with the recommendations of BS 5837:1991. In fact, the only person who should be on site before the arborist is the land surveyor who carries out the topographical survey.

The arboriculturist then confers with the architect, engineer, planner and, if necessary, an ecologist and possibly an archaeologist. From these discussions, and based on the AIS produced by the arborist, a workable layout should emerge which can then be presented to the LPA for informal discussion. The Case Officer will advise where adjustments need to be made, and generally will assist the application if he has been provided with all the data by the design team.

When the format has been agreed as far as possible, a formal application is made and the Case Officer puts the application out to the various consultees. During this process it is quite permissible for the developer's design team to continue discussions with the LPA team.

Finally, the Case Officer consults with the CPO and a report to the Planning Committee is prepared which details the entire history of the application, the results of consultations, comments from the public and interested parties and will include a recommendation that the Committee either *Approve* and grant permission or that they *Defer* the decision pending more information and a visit to the site or, that they *Refuse* the application consent.

If the application is refused the applicant can re-submit within six months having addressed the reasons for refusal or he may appeal against the decision to the Department of the Environment. In the case of an appeal, the Secretary of State appoints an Inspector from the Planning Inspectorate who will decide upon the application having heard all the evidence on each side. The decision of the Secretary of State, through the Inspector, is final (10).

If, however, the application is approved, then there are always conditions attached to the grant of consent, some of which relate to the protection of trees to be retained on the site. These conditions represent the final statutory protection that the trees have during and following development.

Conditions of Planning Consent. Any conditions which are attached to a consent must be reasonable and enforceable and suitable to the individual site (10). Often the conditions which relate to the preservation of trees state simply that trees must be protected by the erection of protective fencing in line with the recommendations of the British Standard 5837:1991. This is the weakness in the system. Conditions are often not phrased properly or are not detailed enough to protect trees adequately. It is for this reason also that planners often view the arboricultural aspects of development with scepticism as it is quite possible to produce an extremely logical and supportive AIS but enforcement can be difficult and problematic and result in a loss of trees.

There is a mechanism within the Planning Act to turn the Conditions into a stronger legal agreement, i.e., Section 106. What this means is that an LPA can bind the developer legally and force him to put right damage at his own expense and or compensate the LPA. Section 106 Agreements (S106) are cumbersome and are of little comfort if mature trees are lost, and thus are not favoured by Planning Authorities.

The way to overcome this problem is to require, by condition, that the developer produce a full and detailed Method Statement for Arboricultural Works. The condition should be phrased such that *no development of any sort, including the erection of site offices etc can take place on site until the Method Statement has been submitted to and agreed in writing by the Chief Planning Officer.*

The Arboricultural Implication Study (AIS) and Method Statements

The Arboricultural Implication Study. It has become obvious that the planning process is logical and that it works. However, in order to be in a position to preserve trees on sites, it is essential that all the arboricultural implications are assessed and detailed at the application stage. This includes collection of all data on all trees on site and, with reference to the British Standard, deciding which trees should be retained and which should be removed (Figure 3). It is then essential that all interfaces between the development and the trees, their root zones and their crowns and branches are assessed and discussed, i.e.:

- The protection of trees within total exclusion zones;
- The location and type of protective fencing;
- The location of the main sewerage and water services in relation to trees;
- The location of all other underground services, i.e., gas, electricity and telecommunications;

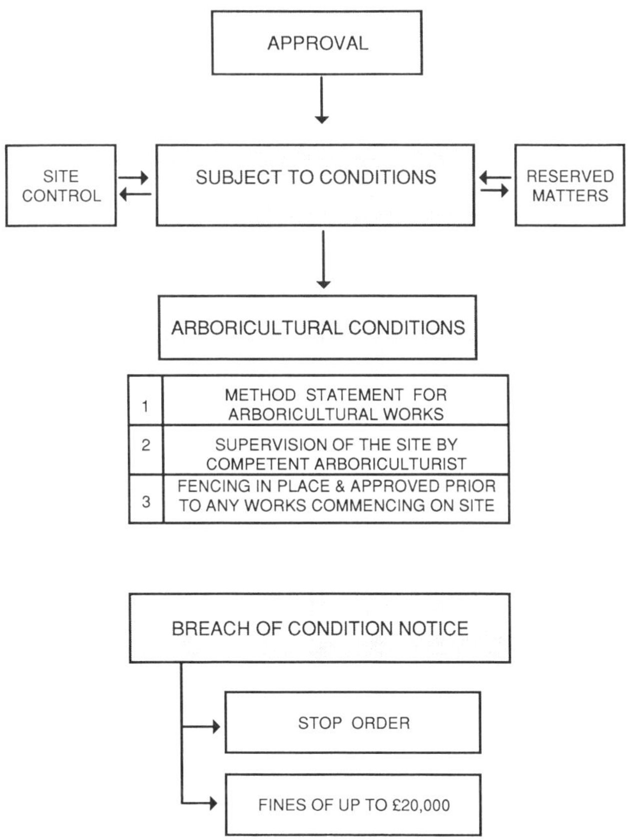

Figure 3. A chart illustrating the implementation of the arboricultural conditions of planning consent on site.

- The locations of roads, pathways, parking and other hard surfaces in relation to tree root zones;
- Provision of design and engineering solutions to the above, for example, thrust boring for service runs; the use of porus surfaces for roads etc and the remedial works to maintain tree health such as irrigation and fertilisation systems; the use of geotextile membranes to control root spread.
- Suggested locations for the site compound, office, parking and site access;
- The replacement plantings necessary to compensate for the necessary losses.

All of the above are presented within the Arboricultural Implication Study prepared in support of the application. The AIS should be supported by relevant drawings of the site and layout with tree positions, the location of fencing and site compounds, means of access, etc., all clearly marked. In addition, the AIS should also contain model planning conditions and a draft form of a Method Statement for Arboricultural Works which would apply to the site.

The information collected, analysed and presented in the AIS format allows the LPA Case Officer to fully assess the arboricultural implications. This detailed analysis allows for the compilation of the final form of the Method Statement.

Method Statements. A draft of the Method Statement (Figure 4) should ideally be contained within the AIS in order that the LPA can comment at an early stage on how it should finally read. The function of the Method Statement is to translate all the necessary arboricultural aspects of the entire development into *Construction Speak* which is readily understood and appreciated by building workers.

It sets out clearly the details of the site and development. It contains details of the Development Team and the Local Authority Team giving names, addresses, telephone, fax and cellular numbers to ensure communication. It includes the arboricultural *Conditions of Planning* and a specification for the type of protective fencing to be used. In addition it contains a Bill of Quantities and Schedule of Works for all tree works on site and indicates when arboricultural advice and guidance is needed during the construction phase. The Method Statement contains a flow chart for use in the site office and sets the chronology of events clearly for both developer and planner.

The success of the Method Statement lies in the necessity of having the site supervised, throughout the development by a competent and qualified arboriculturist. This may seem odd at first glance, however, sites are supervised by architects and engineers who relate to both Building Control Officers and Local Authority Engineers as appropriate. Site agents are used to architects, etc., coming on site and issuing various instructions about the building process. All that is suggested in this paper is that the same discipline

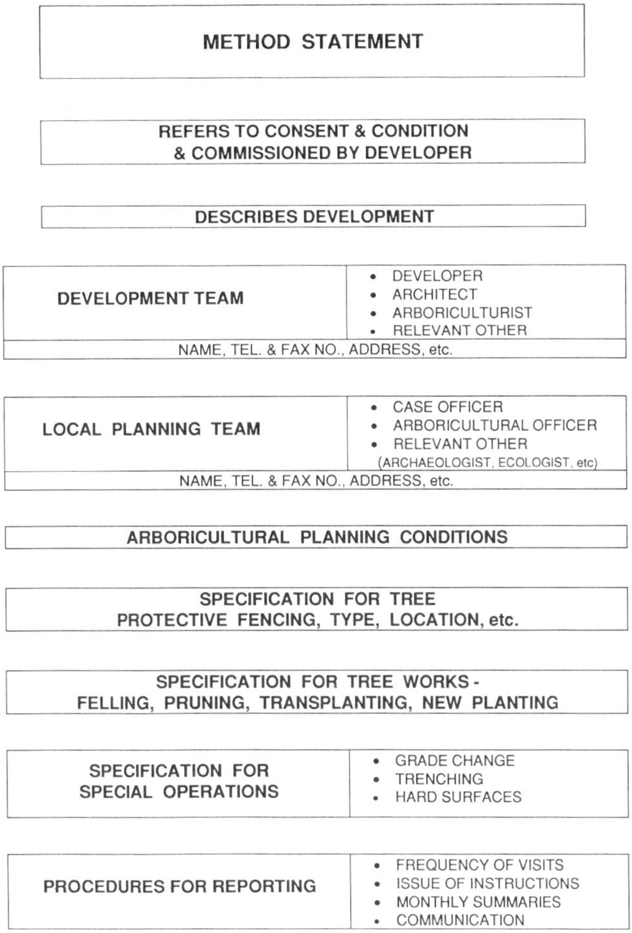

Figure 4. An illustration of the contents and construction of the Method Statement for Arboricultural Works.

is applied to the arboricultural aspects of the development.

This system works well and site agents seem happy to accept that an arboriculturist, "the tree guy" comes on site, inspects the tree aspects with them and writes an *Arboriculturists Instruction* if necessary. The system of operation is similar to that used by architects and engineers on sites and therefore, it fits the construction mentality.

A key element within the Method Statement is the reporting procedure. The developer essentially pays the consulting arboriculturist to keep his site in order, in arboricultural terms, to ensure that all arboricultural conditions of planning consent are enforced and for the arboriculturist to be available to deal with any emergencies which might arise. The arboriculturist completes a site inspection record at each visit and this is counter signed by the site agent. A copy is left on the site and a copy forwarded to the developer's home office. Monthly summaries of site activity are prepared and forwarded to the planning department. Should a planning officer visit the site and notice something untoward, i.e., fencing not in place, then he or she checks with the site agent, examines the inspection records and notes if the arboriculturist has visited and if an instruction has been issued. The next person the planning officer should contact is the consulting arboriculturist, and, thus, the developer is insulated from any problems.

The key is communication. There is an identified methodology and chronology, a sequence of activities and a chain of people with varying degrees of responsibility. By following the routes, most, if not all, arboricultural site problems can be solved before they develop into major incidents. This gives the Local Planning Authority control over the arboricultural aspects of the site, which is unique in arboricultural terms, but quite common in all other aspects of building and development.

The Planning Framework

This system has been in operation in parts of Britain for some years now. Mostly it is enforced by Planning Conditions and is favoured for the following reasons:

- Lines of communication are secured;
- Planning Conditions are more easily enforced;
- Statutory Controls are strengthened;
- Development Sites are rationalised;
- Special features are specified, i.e. foundations, surfaces, protection, etc.;
- Services can be controlled; and
- Site supervision is guaranteed and supplied by the developer.

Some Local Planning Authorities have taken the approach a stage further. They have produced, in cooperation with Consulting Arboriculturists, O'Callaghan Associates Ltd, a set of guidance documents for developers which explains the system clearly and concisely. These are known as Supplementary Planning Guidelines (SPGs) and are part of the Authority's Local or Unitary Development Plans. The documents, once adopted by the Local Council, become Policy Documents, similar to City Ordinances in the United States, and have statutory power. The SPGs include a digitised template pro-forma for assessment of Arboricultural Implication Studies and Method Statements submitted.

Where this has occurred, the Councils have taken time to run workshops and seminars for architects and developers working in their area, on how the system works and how it should be implemented. The result is a cooperative approach; the protection of trees on development sites rather than the more usual "stand off" position which can so often result in acrimony, ill feeling and appeals.

By including this approach in the existing planning framework, the result is that the excellent guidelines for Trees in Relation to Construction, contained within the British Standards are made to work effectively. The main reason being that the approach is a cooperative one, communications are assured and developers have the "tree problem" addressed and removed.

Conclusions

A cooperative approach to the development of land and the preservation of trees has been demonstrated to be possible and effective by experience in England. The system works because it achieves two important goals. First, it is framed in a format that is familiar and user friendly to the developer and construction workers. Second, it fits easily into the existing planning framework in Britain.

The development of this approach relies on the guidelines contained within the current British Standard (2). and the latest Government advice (10). It succeeds because it starts with a full implication study of the arboricultural aspects of the development which provides the planning department with all information needed to frame a decision. It succeeds after approval, by virtue of the Method Statement for Arboricultural Works which, in concept, is no different from other methodologies builders have to work with, for example, a methodology for installation of drains, or construction of foundations. In short, It is user friendly to the construction workers.

Finally, the system succeeds because the site is supervised by a competent and qualified arboriculturist. This provides the Planning Authority with confidence that the conditions of planning consent will be implemented and that they will receive regular reports on arboricultural progress on site. Policing of the site is arranged and paid for by the developer. This is essentially the same process that is employed in all other aspects of the building process.

In terms of benefits to our profession, the arboriculturist is a key player in the design team, has authority on the job and can issue instructions to site agents in the same way architects and engineers do. Because the arboriculturist is the first of the design team on the site, he or she has an influence at the planning stages and can influence the layout such that tree problems are designed out at an early stage. This can be achieved because we have eliminated the concept of the *Tree Survey* and have replaced it with the *Arboricultural Implication Study*. This places the arboriculturist at the same professional level as all other members of the development design team.

The end result is that the trees are retained in healthy condition. Confrontation, acrimony and ill feeling are eliminated from the process, and sites tend not to move from one arboricultural crises to the next. We have shown that it is possible to combine effective and profitable development with the preservation of trees in healthy condition without compromising our integrity.

Acknowledgments

We are grateful to the many planners and arboricultural officers with whom we discussed and developed this approach. Specifically, we acknowledge the assistance and inspiration of Les Round, Trafford Borough Council, and the help and assistance of Mike Mottram and his team at Ryedale District Council. We thank Bellway Homes (NW), Alfred McAlpine Homes (NW) Ltd, Persimmon Homes Ltd and Pembroke Homes Ltd among many developers who have worked with us and benefited from the process.

Glossary of Terms Used

Arboricultural Officer. The Municipal Arborist/Forester, (A.O.), who may or may not be based in the Planning Department.

Case Officer. This is the member of the Planning Department who deals specifically with an application(s) for consent to build etc.

Chief Planning Officer. Head of the Planning Department of a Local Authority, who is responsible to the Elected Council Members and the Chief Executive for administration and implementation of all planning policies.

Committee Report. This is the report written by the Case Officer or CPO for the agenda of the Planning Committee containing analysis of the application and a recommendation for the Committee.

*DoE Circula*r. Guidelines setting out the interpretation successive Secretaries of State have applied to the Planning Acts on a range of issues.

Department of the Environment. DoE, the Central Government Department with responsibility for Local Government and Environmental matters.

Local Authority. The Municipal, County or City Government.

Planning Inspectorate. A division of the DoE which is responsible for administration of Appeals, Public and Local Inquiries, enforcement issues etc.

PPG. Planning Policy Guidance notes. These are the Secretary of State's guidance on a range of general and specific aspects of planning policy.

SPG. Supplementary Planning Guidance notes, Local Planning Authorities policies on a range of general and specific issues, i.e. Trees & Development. They are usually part of the Local or Unitary Development Plans.

SULE. Acronym for Safe Useful Life Expectancy

UDP. Unitary Development Plan which is the Local Planning Authority statement on Policy, Zoning, Greenspace, Heritage, Conservation, Transport, Development, etc., for five years.

Literature Cited

1. Barrell, J.B. 1993. Pre-planning tree surveys: Safe useful life expectancy (SULE) is the natural progression. Arboric.J. 17: 33-46.
2. British Standard 5837:1991. 1991. A guide for trees in relation to construction, BSI Publications, London.
3. Cutler, D.F. 1993. Interactions between tree roots and construction work. Arboric.J. 17: 47-56.
4. Mattheck, C. & H. Breloer. 1994. Field guide for visual tree assessment (VTA). Arboric.J. 18: 1-23.
5. Moore, V. 1987. A Practical Guide to Planning Law, Fourth Edition 1994, Blackstone Press Ltd., London. 415 pp.

6. O'Callaghan, D.P. 1989. Remedial tree work - A long term investment: A long term project. Arboric.J. 13: 303-312.
7. O'Callaghan, D.P. 1991. Legal protection for trees in Britain and Ireland. J. Arboric. 17: 306-307.
8. Planning & Compensation Act, 1991. HMSO London.
9. Town & Country Planning Act, 1990. HMSO London.
10. Tree Preservation Orders Guide. 1994. A guide to the law and good practice. Department of the Environment, London

Case Study, The Enid A. Haupt Garden Soils, Smithsonian Institution, Washington, D. C.

James C. Patterson and Paul H. Lindell

This paper discusses the retrofitting required to ameliorate a poor soil condition in the Enid A. Haupt Garden at the Smithsonian Institution. The purpose is to stress the significance of specifying an appropriate soil environment for a large pallet of specimen horticultural plants and assuring professional overview during construction. Plants were selected in nurseries and other sources at a high cost to create an outstanding formally landscaped garden above two underground museums. Unfortunately, there was little direct control exercised by the contracting agency over the quality and placement of the topsoil. Thus, there were plant decline symptoms and failures that were traced to the poor soil conditions. This paper will address the construction of the gardens, source of the soil and its characterization, and the significant efforts since construction to modify the system to support the landscaping.

The garden is located in the southeast plaza of the Smithsonian Institution's Castle, the first Smithsonian building, between the Castle and Independence Ave. SW and occupies 4.2 acres (1.6 hectares). The history of the site indicates that it had been exposed to many diverse uses. It has been the location of a railroad facility, buffalo and stock pens, parking facilities, research laboratories, the National Air and Space Museum, and other intense uses prior to its selection for the Garden.

Enid A. Haupt, a noted horticultural philanthropist, donated the money to acquire plants and create a formal landscaped garden. The design concept was to create a garden which would provide a place for contemplation and an escape from the busy National Mall. Large specimen plants were purchased from east coast nurseries to provide instant effect at the request of the donor. Construction began in 1984 and was completed in 1988.

On the NE corner of the site was a very large Greenleaf linden (*Tilia platyphyllos 'vitifolia'*) which was to be preserved during site construction and was to become a focal point in the Garden. Efforts to retain this tree included installation of a misting system in the crown, lightning protection, pizometers to monitor soil moisture levels in the rooting zone during the 60 foot excavation and dewatering of the site, and the careful watchful eye of the Institution's Horticulture Services Division. The tree was protected in an 80 x 80 ft fenced exclosure. During construction there was a noticeable thinning

---

Research Agronomist, Center for Urban Ecology, National Biological Service, and Landscape Architect, Smithsonian Institute, Washington, D. C.

of the trees' crown. Also, there was an awareness of some trunk decay on the SW side of the trunk prior to construction. In any event, the decision was made to try to preserve the tree particularly since the species was uncommon in the Washington, D.C. area. However, even though heroic efforts were made, it succumbed to natural forces and collapsed two years after construction.

The two underground museums, the National Museum of African Art and the Arthur M. Sackler Gallery, were constructed and the garden established on the roofs of these museums. Site excavation occurred to depths of about 60 feet (about 18.3 meters) below the existing surface. Site dewatering and removal of soil materials was substantial.

The primary contractor was enterprising and stockpiled the sandy subsoil deposits at his business location. Concurrently, he was involved with dredging sediments from a facility along the Potomac River in Alexandria, VA. The dredged material was dewatered and saved. He engaged an agronomic consultant to evaluate the feasibility of combining these two soil materials to see if the mixture would meet the General Service Administration's (GSA) specification for the "topsoil" for the Garden. GSA acted as the contracting office for the project. The consultants stressed that with proper blending techniques and appropriate proportions of ingredients, the mixture would be acceptable. However, close control over the mixing of the soil materials was vital. GSA, even though professional consultation was available, decided this overview to be unnecessary. The resultant soil mixture was quite heterogeneous, had an excessive pH, between 7.0 and 7.8, and the soluble salt content exceeded 500 parts per million (ppm). Both of these soil parameters significantly exceeded specifications.

It is important to note that the data presented in this paper are taken from several consultant reports as well as the authors information. As a result there will be field sampling, procedural, and sample handling differences between different soil testing laboratories. This is the major reason for the occasional wide differences in observed or cited data. However, the data are reflective of the poor soil conditions of the site. Normally, it is always advisable to have samples tested using the same laboratory and laboratory techniques for direct comparisons over time.

As stated, and regretfully so, the GSA deemed the mixture acceptable. Further, the proposed soil depth was to be 6 to 8 feet (1.8 to 2.0 meters) but was reduced by GSA to 16 to 18 inches (42 to 46 cm) at many locations while the deeper depths existed only at a few locations. A summary statement by the agronomic consultant reflects the importance of proper handling of this situation, "since the topsoil mixture potentially could result in anaerobic conditions and that reduced drainage rates might be a problem because of the high biological oxygen demand and slow oxygen diffusion at depth, serious concerns are expressed regarding the mixture (14)."

As background, key portions of the GSA specifications for the Garden are provided. Unfortunately the specifications allowed for considerable flexibility of source materials for the soil and no on-site professional control of daily operations.

1. Composition: Topsoil shall be fertile, natural friable soil suitable for growth of plants. It shall contain considerable decomposed organic matter, finely divided and readily discernable by visual inspection, but in no case less than 2% by weight.
2. -Soil acidity range from pH 5.5 to 7.0 inclusive.
   -Soil salinity by measure of soluble salts not to exceed 500 parts per million.
   -Soil fertility rated "high" in natural nutrients.
   -Topsoil shall be classified as a loam as defined by the USDA-SCS classification system.
   -Permeability shall be approximately .0001 cm/sec, based upon grain size (ASSHTO-T88-78).
3. The contractor was directed to provide a proposed system to amend the soil to insure that it provided a pH between 6.1 and 6.5 by adding such materials as limestone or aluminum sulfate.

To further complicate on-site conditions, placement of the material occurred with the first load placed and each successive load was maneuvered over the previous load thus resulting in excessive compaction. The result of inadequate mixing of ingredients and this type of placement was a heavily compacted heterogeneous soil mixture. After placement, heavy equipment and vehicles continued an unrestricted use of the site, further complicating the soil situation. The final soil depths were variable and ranged from about 16 inches (40 cm) to about 6 feet (183 cm).

As a result, significantly poor drainage conditions occurred and the irrigation system required major corrective maintenance. The literature provides a clue as to how much time is required for natural recovery of a compacted soil condition, if left untouched by human activity. The estimate is that "soil recovery, based on a linear model using bulk density values, suggests that approximately 100 to 130 years are required for complete loosening to occur for abandoned roads, and that 100 or fewer years are necessary for complete amelioration of the foundation periphery areas (7)." This indicates the long-term effects of soil compaction and reinforces the necessity to provide suitable soil and minimize compaction during construction.

The sandy subsoil salvaged from the excavation was a mixture of ungraded sand which easily compacted. The agronomic consultant sampled the sandy subsoil and confirmed that little of the desirable very coarse sand fraction existed (Table 1). A pH and soluble salt analyses of selected soil mixtures were conducted to provide an initial evaluation of the anticipated soil chemistry (Table 2). These data confirmed the existence of poor soil chemistry of the mixed or blended materials. Professional overview could likely have prevented this occurrence had it been available. Soluble salt

Table 1. Textural classification of field-blended topsoil material* (pre construction)

Size fraction	Percent (wgt)
v. coarse	2.2
coarse	9.6
medium	13.9
fine	7.9
v. fine	6.4
total sand	40.0
silt	45.0
clay	15.0

Textural class: loam
*Source: Geo-Sci Consultants Inc., March 28, 1985.

Table 2. Select soil test results* (Data represent select soil mixtures from contractor stock piles)

Sample	Soluble salts (mmhos/cm)	Soluble salts (ppm)	Sodium (ppm)	Calcium (ppm)	pH
1	2.8	1792	48	1360	7.0
2	4.4	2816	46	1850	7.3
3	3.6	2304	47	1680	7.3

* A&L Agricultural Laboratories, September 21, 1985.

levels were quite high, 1360 to 1850 ppm, while the pH range was 7.0 to 7.3 (14). It is important to note that these data are from prepared mixtures of individual ingredients, subsoil and sandy dredge spoil, presumably from contractor stockpiles.

Soil Description

Problems were encountered with the salvaged soil mixture early during transplanting. The dredge spoil materials were inappropriately mixed with the sandy deposits in an attempt to provide the specified "loam" topsoil mixture. This mixing operation was apparently accomplished away from the Garden site. Likewise, we observed that a buffered-alkaline composted sewage sludge product with a high soluble salt content had been blended with the sandy material and dredge spoil. The resultant mixture provided the heterogeneous compacted soil mantle with both a very high pH and soluble salt content (Tables 3 - 6).

Analysis of the soil confirmed the variability. Results from soil chemical

Table 3. Soil test results - Chemical properties*(post-construction)

Depth (in)	Organic matter (%)	Cation exchange capacity (meq/100g)	pH	Soluble salts (ppm)
Location 1				
0-3	1.9	12.4	7.3	500
13-15	2.2	19.5	7.8	595
17-20	0.3	4.4	7.7	<119
Location 2				
0-3	2.6	14.9	7.6	238
10-12	2.5	14.6	7.8	214
22-25	0.2	2.8	7.7	<119
Location 3				
0-2	2.8	16.8	7.5	833
6-10	1.6	16.3	7.9	333
17-19	0.2	2.5	7.5	<119
Location 4				
0-5	2.5	13.1	7.5	547
12-20	2.2	14.6	7.6	381

*Source: Geo-Sci Consultants, Inc., September 9, 1988.

Table 4. Soil physical properties* (post construction)

Depth (in)	Bulk density (g/cc)	Porosity (%)	Field moisture (%)	Hydraulic conduct. (iph)	Permeability class
Loc. 1					
13-15(A)	1.61	39.2	20	0.007	v. slow
17-20(C)	1.60	39.6	6	6.000	rapid
Loc. 2					
10-12(A)	1.52	42.8	23	0.003	v. slow
22-25(C)	1.76	33.8	6	6.950	rapid
Loc. 3					
6-10(A)	1.86	29.8	12	0.032	v. slow
17-19(C)	1.72	34.9	4	5.810	rapid
Loc. 4					
12-20(A)	1.66	37.4	19	0.006	v. slow
Means					
(A)	1.66	37.3	18	0.013	v. slow
(C)	1.69	36.1	5	6.250	rapid

*Source: Geo-Sci Consultants, Inc., September 9, 1988

Table 5. Soil particle size distribution* (post construction)

Depth (in)	Soil particle distribution (%) Sand	Silt	Clay	Textural class (USDA)
Location 1				
0-3	55	33	12	gravelly sandy loam
13-15	44	4	15	gravelly loam
17-20	91	7	2	coarse sand
Location 2				
0-3	45	40	15	loam
10-12	42	41	17	loam
22-25	92	7	1	coarse sand
Location 3				
0-2	46	39	15	gravelly loam
6-10	56	30	14	gravelly sandy loam
17-19	92	6	2	gravelly coarse sand
Location 4				
0-5	50	36	14	loam
12-20	54	32	14	sandy loam

*Source: Geo-Sci Consultants, Inc., September 9, 1988.

Table 6. Sand fractionation* (post construction)

Depth	Sand fractions (%)					Total (%)
(in)	VCS	CS	MS	FS	VFS	
Location 1						
0-3	3.4	13.7	21.9	10.0	6.0	55
13-15	3.1	9.9	14.4	9.3	7.3	44
17-20	4.8	30.7	41.1	11.2	3.1	91
Location 2						
0-3	2.3	10.1	14.4	10.1	8.0	45
10-12	1.8	9.7	14.1	9.2	7.2	42
22-25	5.5	29.0	41.0	13.1	3.4	92
Location 3						
0-2	3.6	11.1	14.0	10.2	7.1	46
6-10	4.7	14.0	18.7	11.8	6.9	56
17-19	9.1	34.5	36.1	9.5	2.8	92
Location 4						
0-5	3.9	12.7	16.6	10.2	6.6	50
12-20	3.8	14.9	18.7	10.4	6.2	54

*Source: Geo-Sci Consultants, Inc., September 9, 1988.

tests for four selected soil pits are provided in Table 3. Note the differences that are apparent between the first two horizons and the lower or sandy horizon. This contrast is consistent for Tables 3 - 6 for these sampling locations and suggests the presence of a significant lithologic discontinuity. Discontinuities of this nature may significantly alter soil physical characteristics such as has been frequently detailed in our urban soil research (3,4,9,10). Likewise, field observations by the authors at several other soil pits verify that, because of non-uniform mixing of the soil ingredients, lithologic discontinuities occur randomly within the soil mantle. Of particular concern is the pH which remains high, pH 7.3 to 7.9. The soluble salt content (Table 3) remains elevated but it is encouraging to see that these results are below the four digit levels noted in Table 2 for the preconstruction prepared soil mixtures. The organic matter content was moderate and the cation exchange capacity (CEC) acceptable for the surface horizons but understandably low for the deeper sandy horizons (14).

The soil physical parameters (Table 4) indicate that significant compaction remains which may inhibit root extension (compare Table 4 and Table 8). Part of the reason for the differences in densities is likely because of the analytical procedures involved, Geo-Sci Consultants, Inc. (14) used the clod method while Law Engineers used a neutron surface density probe (8)(Table 7). Considering that the literature provides an indication of critical bulk density for normal root extension to be about 1.55 g/cc for trees, none of the sampling locations fell below this parameter (Table 10). It is important to note that the density data in Table 4 indicate high soil densities at depths as well. The Law Engineering densities expressed in Table 8 are near-surface observations because of the neutron method used. It is evident that the entire soil profile remains very dense while porosities and total pore space remains low. Porosity evaluations in Table 4 can be compared to the calculated

Table 7. Select soil chemistry data* (post construction)

Location	Calcium (ppm)	Soluble salts (ppm)	pH 1988	pH 1995**	Organic matter (%)
11	132	900	7.1	5.9	5.0
23	99	700	7.1	6.8	4.7
25	202	1200	7.1	6.3	4.9
27	142	900	7.0	6.7	4.0
36	174	1000	7.0	5.7	4.9
52	299	1400	7.0	6.7	4.8
62	431	1800	7.0	6.6	6.1
78	46	400	7.0	6.2	4.4

* Source: Law Engineering, December 16, 1988.

** 1995 data conducted by the University of Maryland Soil Testing Laboratory.

Table 8. Soil physical condition* (Post construction)

Sampling location	Total density pcf***	Bulk density g/cc	Calc. pore total space (%)**
1	120.4	1.93	27.17
2	124.3	1.99	24.91
3	120.9	1.94	26.79
4	125.0	2.00	24.53
5	109.6	1.75	33.96
6	105.8	1.69	36.23
7	120.3	1.93	27.17
8	122.1	1.96	26.04
9	120.6	1.93	27.17
10	121.2	1.94	26.79

* Source: Law Engineering, December 16, 1988.
**A soil density of greater than 1.55 g/cc or a total percent pore space of less that 35% have been demonstrated some researchers to be hostile environments for normal root growth.
***pcf represents pounds per cubic foot, and engineering term for density.

values in Table 8. The porosity data indicate that root elongation could be restricted (compare Tables 8 & 10). None of the sites observed by Law (Table 8) has porosities greater than 40% (8). Even if we consider 35% total pore space, as other research indicates is the critical level, only one of the observed sites is acceptable. Table 11 provides information on the density and porosity relationships of common construction materials for comparative and illustrative purposes.

Field moisture, hydraulic conductivity, and permeability classification indicate that considerable improvement is still required to achieve some semblance of normal plant root growth (Table 4). The very low hydraulic conductivity and very slow permeability classifications dictate that the site manager will need to closely monitor soil moisture conditions to prevent saturation and/or desiccation.

Soil particle size (Table 5) and sand fractionation data (Table 6) support the existence of the heterogeneous soil mixture and presence of lithologic discontinuities. Note the very low percentage of very coarse sand present in the sand fraction. It is this coarse sand that would assist with improving soil physical relationships if it were present. However, it is important to realize that very high volume percentage of coarse angular sand is required to improve soil physical relationships (10).

Law Engineering (Table 8) provided soil chemistry data which support the data of Table 3 (8). Table 8 provides an opportunity to compare the soil pH data of 1988 and recent results from 1995 and the University of Maryland Soil Testing Laboratory. These data support the philosophy that progress has been made toward a more favorable soil chemical condition.

Table 9. Selected soil chemistry data* (Post construction)

Depth	Acceptable Turf				Dead Turf			
	Sample	pH	EC	SS	Sample	pH	EC	SS
in.			mmhos/cm	ppm			mmhos/cm	ppm
0-1.25	2a	6.20	0.47	470	1a	3.50	2.80	2800
1.25-3	2b	7.00	0.42	420	1b	7.05	1.10	1000
0-1.25	4a	5.18	0.49	490	3a	2.43	2.86	2860
1.25-3	4b	7.03	0.21	210	3b	6.97	1.06	1060
0-1.25	6a	5.10	0.49	490	5a	2.43	1.75	1750
1.25-3	6b	6.82	0.52	520	5b	6.47	0.96	960
0-1.25	8a	5.88	0.33	330	7a	1.88	5.53	5530
1.25-3	8b	6.94	0.51	510	7b	6.05	1.20	1200

*Source: Weil, Dr. Ray R., March 26, 1989.
NOTE: Salt levels are reported as electrical conductivity (E.C.) in mmhos/cm. and can be converted to the approximate equivalent in ppm by multiplying by 1000.

Table 9 provides pH values of selected turfgrass site conditions within the Garden after receiving a soil surface treatment with sulfur to lower the soil pH. This application occurred after Dr. Weil's (15) first site assessment which determined that the surface 0 to 3.2 cm and the 3.2 to 7.6 cm soil depths contained contrasting conditions as related to pH and soluble salts. Between his visits, an application of sulfur was made in an attempt to lower the pH of these surface soils. However, the sampled areas of acceptable vs dead turfgrass yielded very interesting results and that the downward movement of the sulfur was very restricted. The contrast of these soil depths was dramatic. The apparent downward movement of sulfur was prevented by the lithologic discontinuity created at the surface or sod layer. Additional analysis of the presence of this layered condition is necessary to determine if the aeration practices used have improved conditions. A partial explanation of the condition may be that a lithologic discontinuity exists between the sod layer and the underlying soil.

Recommendations

In 1988, there were questions raised about how the soil problems could be corrected and plant vigor and appearance enhanced. Unfortunately, the poor soil conditions had occurred and could have easily been corrected by supporting professional agronomic oversight of the soil mixing and placement during construction. As a result only heroic efforts would begin to generate improvement.

Table 10. Critical bulk density levels for tree and turf survival (density above these levels and porosity below these levels plants have difficulty surviving).

Literature citation	Plant type	Bulk density (g/cc)	Pore space (calc %*)
-Bertrand & Kohnke, 1957	grass	1.50	43.40
-Daniel, 1978	turf	1.31	50.57
-Sills & Carrow, 1983	turf	1.34	49.43
-Carrow et al., 1992	turf	1.34	49.43
-Forrestall &	trees	1.30	50.94
-Smiley & et al., 1990	trees	1.50	43.40
-Craul, 1992	trees	1.55	41.45
-Craul & Patterson, 1992	trees	1.55	41.45
-Smiley, 1992	trees	1.50	43.40
-Smiley &et al., 1990	trees	1.50	43.40

*Note: the percent total pore space was calculated using 2.65 g/cc as the particle density.

Summary: For turf: The literature indicates that the soil density should not exceed 75.9 pcf or 1.37 g/cc and the total pore space should be less than 48.21%. (NOTE: these data are the averages of the four literature citations for turf.) For trees: The literature indicates the soil density should not exceed 86.3 pcf or 1.48 g/cc and the total pore space should be less than 44.13%. (NOTE: these data are the averages of the five literature citations for trees.)

Table 11. Physical characteristics of four construction materials*

Material	No. of samples	Bulk density range g/cc	Bulk density ave. g/cc	Particle density g/cc	Calculated pore space %
cinder block	3	1.67-1.71	1.70	2.64	35.6
brick	3	1.43-2.01	1.75	2.72	35.7
asphalt	3	2.17-2.22	2.19	2.35	6.8
concrete	3	2.23-2.27	2.26	2.47	8.5

*Patterson, 1976.

Recommendations proposed and implemented:

1. Remove the entire planting and soil and import or provide a suitable soil. This option was eliminated because of its drastic nature and high cost.
2. Soil improvement program was initiated in1988 and included:

   a. Provide vertical aeration and vertical mulching. Use hand or power augers and/or high pressure air/water/amendment injection methods.

   b. Follow the vertical mulching applications using a light-weight aggregate and an organic matter backfill to maintain holes that are open to the soil surface. Follow this with sulfur additions to gradually amend the pH.

Routine checking of soil pH and soil chemistry will be necessary to track progress.

c. Enhance irrigation once aeration, mulching, and sulfur applications are completed to encourage infiltration. This will improve leaching of excess salts and pH.

d. Provide and encourage earthworm growth in the soils. Two species were recommended: *Lumbricus terrestris*, a burrowing species, and *L. longa*, a species which can be cast on the soil surface.

e. Begin a vigorous and continuous vertical mulching maintenance regime using organics and light-weight aggregates.

f. Continue periodic site evaluations of the soil conditions to track progress.

Conclusions

In general, the Garden is making a slow but encouraging recovery. The success will require cyclic monitoring of field treatment activities and perhaps altering the recommended methods when new research provides improved and effective measures to deal with such difficult situations. Plants have responded favorably to the current amendment procedures but, as the data indicate, there is work remaining. The best and most efficient method is to have complete control over all site and construction activities from initiation of the project so that favorable soil conditions are provided.

Literature Cited

1. Bertrand, A. R., and H. Kohnke, H. 1957. Soil Science Society of America Proceedings. 21: 45-48.
2. Carrow, R. N., and A. M. Petrovic. 1992. Effects of traffic on turfgrass. **In** Turfgrass, Agronomy Monograph no. 12, ASA-CSSA-SSSA, 677 S. Segoe Rd. Madison, WI 53711. pp 285-329.
3. Craul, P. J. 1992. Urban Soil in Landscape Design, John Wiley & Son, Inc. 396 pp.
4. Craul, P. J., and J. C. Patterson. 1992. The urban soil as a rooting environment, **In** Proceedings of the Fourth Urban Forestry Conference, St. Louis, MO., pp 97-102.
5. Daniel, W. H. 1978. The science of soils from New Turf Handbook. Weeds, Trees, & Turf. 17(1): 26, 28, 30, 32, & 34.
6. Forristall, F. F., and Gissel, S. P. 1955. Soil Science Society of America Proceedings. 19: 384-389.
7. Knapp, P. A. 1992. Soil loosening processes following abandonment of two arid western Nevada townsites. Great Basin Naturalist. 52(2): 149-154.
8. Law Engineering. Dec., 1988. Topsoil Investigation. 4460 Brookfield Dr., Chantilly, VA. 22021.
9. Patterson, J. C. 1976. Soil compaction and its effects upon urban vegetation. **In** Better trees for metropolitan landscapes symposium proceedings. USDA Forest Serv. Gen. Tech. Rep. NE-22:91-102.

10. Patterson, J. C., and C. J. Bates. 1994. Long term, light-weight aggregate performance as soil amendments. **In** The Landscape Below Ground, Proceedings of an international workshop on tree root development in urban soils. International Society of Arboriculture. pp. 149-156.
11. Sills, M. J., and R. N. Carrow. 1983. Turfgrass growth, N use, and water use under soil compaction and N fertilization. Agronomy Journal. 75: 488-492.
12. Smiley, E. T. 1992. Growth response of trees treated with soil aeration machines. **In** Growing Points, Univ. of California Coop Extension Service: 29(4): 3-4.
13. Smiley, E. T., G. W. Watson, B. R. Fraedrich, and D. C. Booth. 1990. Evaluation of soil aeration equipment. Journal of Arboriculture. 18(5): 118-123.
14. Wagner, D. P., and P. A. Snow. Oct., 1984-Sept., 1988. Geo-Sci Consultants, Inc., P.O. Box 1125, College Park, MD. 20740.
15. Weil, R. R. July, 1987-Mar., 1989. Soil Scientist/Agronomist. Hyattsville, MD.

Parkway Tree Protection Program

T. L. Green and P. Young

The Village of Glen Ellyn is a forested suburban community located 30 miles west of Chicago, Illinois. In 1987, Hill Avenue, a major arterial street with its parkway containing a mature tree canopy, was slated for widening, curb and gutter repair and other utility improvement. The work was scheduled for two phases. In Phase I (1988) the Village elected to use a concrete-slip form paving machine to install curbs and gutters along with new pavement. Approximately four feet of parkway space outside the paving machine was excavated to accommodate the machine's tracks. This caused a greater loss of root zone for the parkway trees than that expected by residents living on Hill Avenue. Few protection measures were taken for the parkway trees before construction. Some protection measures were implemented after construction work started. Glen Ellyn residents have a great regard for their trees and were concerned over possible injury to the parkway trees as a result of construction during Phase I. There was a change in construction tree-protection procedures for Phase II (1989). The design contractor, working with Morton Arboretum personnel, Village officials and residents developed a 14-point Tree Protection Program. This program is considered to be a model program for protection of parkway trees during municipal street and / or utility improvement projects.

Hill Avenue: Phase 1

In 1988, work was begun to widen the west end of a major arterial street, Hill Avenue, in Glen Ellyn. In addition to widening, Hill Avenue was to receive new curbs and gutters and new storm sewers. Also planned was repair of water, sewer and gas utilities where needed. No provisions had been made in the initial planning or drawing of construction details for the protection of trees growing in the parkway space. Detailed drawings, however, did contain locations and diameter measurements for existing trees. Although the Village of Glen Ellyn had requested assistance from the Morton Arboretum to offer guidance for protection of the trees for Phase I, the provisions necessary to protect trees were not written into the original contract. Thus, little could be done.

The residents of the west end of Hill Avenue did not realize what was in

The Morton Arboretum, Lisle, IL 60532. Dr. Green's current address is 304 Knoblauch Hall, Western Illinois University, Macomb, IL 61455
Peggy Young is Forestry Superintendent, Village of Glen Ellyn, 30 South Lambert, Glen Ellyn, IL 60137

store for them or their trees until after construction began. The installation of curbs and gutters, where there are adjacent trees, can be done by hand digging and forming within one foot [30 cm] of the pavement surface. However, for Phase I a paving machine was selected that was capable of laying pavement, curbs and gutters all at the same time. From the engineering viewpoint, it produces fewer seams and also reduces long term maintenance. This machine could be used without increasing the initial cost. In order to operate this machine, however, a gravel base must be laid to accommodate the caterpillar-like tracks on which it moves. To provide the gravel base an additional 4 ft [1.3 m] of soil beyond the pavement on each side of the street needed to be removed. Since space for roots in the parkway is usually at a premium, this track base resulted in a significant root loss for many trees that were considered to be mature or over-mature.

A base for the new pavement also had to be prepared. This was done by removal of existing pavement and subsoil. Some subsoil was suitable for reuse. It was stored on the adjacent parkway. This resulted in the burial of parkway tree root systems with soil high in clay content for approximately 6 months.

Subpavement and parkway soil were removed with excavating equipment. As equipment encountered lateral roots of parkway trees, the roots were physically torn from the trees. As the trucks dumped their loads of gravel, the top of the truck's bed rose over 20 ft [6 m] in the air and struck tree branches as it drove down the street. Striking large branches with a raised bed caused damage to the trucks as well as to the trees. The area beneath some parkway trees was used for storage of heavy construction hardware. For example, the large heavy metal collars that hold manhole covers were seen leaning against trees.

Some of the sidewalks had to be replaced. The replacement process required breaking the old sidewalk into pieces, picking up the pieces, preparing a new sidewalk base, building forms and pouring new concrete. In the sidewalk replacement operation, it was the pickup of the pieces and preparation of a new sidewalk base that were the most injurious to the parkway trees. End loaders were used to pick up broken sidewalk. As they maneuvered around the parkway trees, many trees were inadvertently struck. Removal of the old sidewalk base with buckets often ripped the lateral roots.

The specifications for the re-grading and re-sodding of the parkway at the completion of the project permitted the use of the high clay content subsoil to be used for backfill. The clay was unsuitable for root survival and growth.

As a final insult for the trees, the Midwest experienced one of the worst droughts in history during the summer and fall of 1988. There were no provisions for watering any of the trees during the construction period. In addition, the Village of Glen Ellyn prohibited its residents from watering their trees (or lawns) either on private or public property.

Hill Avenue: Phase II

The residents living on the east end of Hill Avenue had one year to observe the activities to their west. The failure to protect trees during Phase I was of great concern to Glen Ellyn residents and precipitated considerable resident-municipal government interaction. Glen Ellyn government officials brought together the design engineering contractor, Donohue & Associates, and Morton Arboretum staff to develop a comprehensive tree protection plan. A 14-point program was then presented by the design engineers at a public hearing allowing residents to question the proposed project. This 14-point program helped correct deficiencies in Phase I. This was a model set of guidelines which will work for municipalities with or without the services of a professional forester. A 15th point was added to the original after the completion of Phase II.

Tree Protection Guidelines

1. Plan review by arborists. When a capital improvement program is proposed, the first question to be asked should be "Are there any municipally-owned or privately-owned trees within the proposed construction limits of the project?" If so, a person specifically responsible for the administration of tree protection guidelines should be identified (on-site arborist). If the municipality does not have a forester on staff, a qualified consulting arborist should be contracted. The role of the on-site arborist is to review all aspects pertaining to preservation of municipal trees. This consultant should be selected by the municipal forester or other municipal official and **not** by the general contractor or any subcontractor. The fee for hiring the consulting arborist should be independent of the general contractor's fee. If a municipality has difficulty finding a qualified consulting arborist, the International Society of Arboriculture (ISA) can be a source for names of qualified arborists in the area (P.O. Box GG, Savoy, IL 61874). Communities with in-house municipal foresters may not need to hire an outside consulting arborist.

2. Pre-construction tree inventory. The on-site arborist will conduct a pre-construction tree inventory. The tree inventory will determine the following information about each tree within the proposed construction limits: exact location, species, d.b.h., relative health, appraised value, maintenance needs (pruning, fertilizing, cabling, etc.), worthiness for protection measures, consideration for removal where necessary, whether public or private ownership, and other special tree concerns. Protection measures add additional expense. The appraised value of a tree should be equal to the protection expenses. Those trees not worth that expense usually do not need to be protected. Trees thought to be liabilities should be given a separate hazard assessment with documentation and should be removed prior to construction (2).

3. Pre-construction pruning. During the inventory, all trees with large dead branches should be pruned to make the construction site safe for all construction personnel. Some trees may require lower-branch pruning to allow for the operation of construction equipment. It may be necessary to prune to a height of 16' [5.3 m] or more to allow clearance of construction vehicles. Prohibiting the use of large dump trucks may be necessary in certain segments of the construction area. Trees on private property which overhang public property may also need to be properly pruned. Sometimes branches can be tied back to preserve the form of the tree.

Pre-construction pruning should be done by trained personnel. In communities lacking trained forestry personnel, commercial arborists can be contracted. The work performed by the pruning crews should be under the supervision of the consulting arborist.

4. On-site arborist during construction. Most construction companies lack professionally trained arborists. In general, construction supervisory personnel do not have training in arboriculture. Therefore, it is essential to have a qualified arborist who can make on-the-spot decisions. The on-site arborist does not have to be present every minute of the construction operation, but should be accessible through pagers or a remote phone in order to respond to problems requiring immediate decisions for public trees. The on-site arborist should schedule weekly planning meetings with the construction superintendent to discuss proposed construction activities and their potential impact on the public trees.

5. Construction personnel training. All contractor employees, including supervisory personnel, all subcontractors and their personnel, and all part-time and piece-work personnel will be made aware that the trees are valuable and need to be protected. Before anyone can work within the construction limits, they are required to view a video tape which addresses construction damage to trees. The Illinois Arborist Association has produced an video tape (5) that specifically addresses how construction adversely affects trees. Enforcement of the video tape review will be by the on-site arborist. This is easily monitored by attaching a unique sticker to each worker's hardhat. Anyone found on the construction site without a sticker will be sent to a location where the video can be seen. Since some workers may come to the job site only once for a few minutes, e.g., truck drivers who deliver equipment or construction material, it may be necessary to send a video to their company headquarters.

6. Pre-construction root pruning. Preventing tree root damage requires an understanding of root biology. In general, most tree roots are located in the upper 12-24 in [30 - 70 cm] of soil. Roots of trees quite commonly grow farther laterally than the tree height. Few roots grow beneath paved streets. Roots are more likely to follow utility corridors and to grow in the parkway between the street and the sidewalk. Roots will grow beneath sidewalks and to some extent beneath driveways. Roots grow where the soil is reasonably uncompacted and where there is oxygen. When roots are severed, they regenerate near the cut surface, often by producing dense clusters of fine roots.

Construction activity associated with street widening and sidewalk removal adversely affects tree roots. Heavy equipment used to remove broken street and sidewalk pavement compacts soil and reduces oxygen levels. Buckets and blades used to scrape soil may snag lateral roots and rip them apart, often tearing them back toward the tree base. To prevent the ripping, tree roots should be pruned 6-12 in [15 - 30 cm] closer to the tree than the construction limit (Figure 1). This allows for root regeneration during the construction period. If the cut is made at the construction limit, the cut surfaces of the roots remain exposed to air which causes dieback. Large circular saws used to cut concrete provide the cleanest and least destructive severing of underground roots. Specify the use of equipment manufactured by the arborist industry for root pruning. Where underground utilities (e.g. gas, water, electric) are present, hand digging may be necessary. Exposed roots can be cut with pruning saws and chain saws.

7. Tree protection zone. Construction can damage trees in several ways. Lowering grade and installation of utilities severs roots. Raising grade and stockpiling fill or construction material can suffocate root systems. Operating heavy equipment in the parkway compacts the soil, reducing oxygen levels leading to root death. Large equipment used in construction damages bark and branches by physically hitting them.

To prevent construction damage a highly visible physical barrier (e.g. orange plastic snow fence) should be constructed 360° around each tree designated for protection. Single-strand wires or plastic flagging is not considered as an acceptable barrier. Leaving the fence open on one side is not acceptable. Entry within the "non-violation" zone should be permitted **only** under the supervision of the on-site arborist. During the drafting of the initial construction specifications, a pre-set monetary fine should be estab-

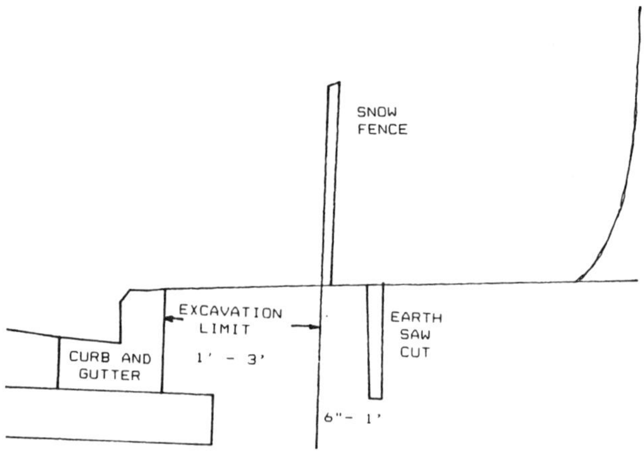

Figure 1. Typical parkway section

Figure 2. Parkway Augering Specifications

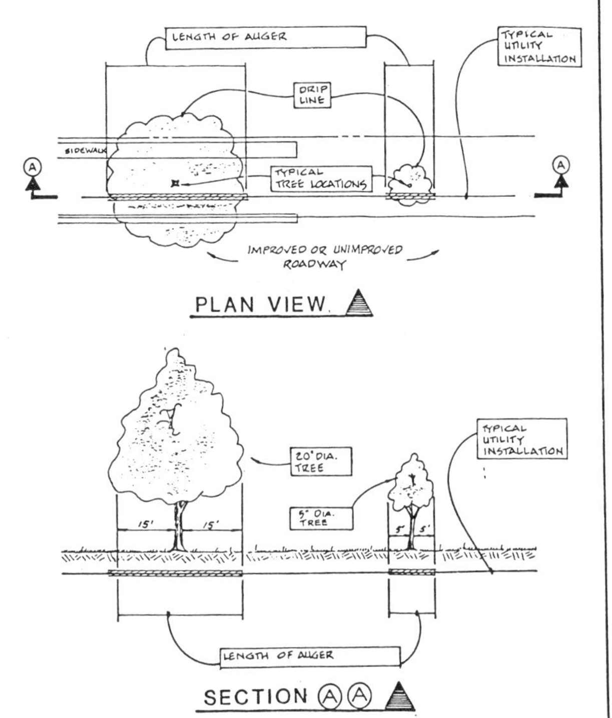

Minimum distance (ft) [m] from tree to trench

Tree diameter (dbh)		Auger distance from tree (radius)	
in	cm	ft	m
0 - 2	0 - 5	1	0.3
2 - 4	7 - 10	2	0.7
5 - 9	12 - 22	5	1.7
10 - 14	25 - 35	10	3.3
15 - 19	37 - 47	12	4.0
19+	47+	15	5.0

Figure 2. Parkway augering specifications

lished for anyone who enters the "non-violation" zone without permission. The fine is very important. Without it, there is nothing to prevent construction personnel from moving the fence limits or from temporarily removing the fence and working around the tree.

There are several ways to establish a minimum "non-violation" zone for parkway trees. The dripline, i.e., branch spread, works well for smaller trees. For large spreading trees, some arborists use a 12 in [30 cm] radius for each inch [2.5 cm] of tree diameter. Where augering (see #8) is recommended, guidelines have been determined for minimum distance for trenching that can also be used for minimum fence radius (3).

8. Utility augering. Since most tree roots are located near the surface, trenching for the installation of utilities can cause significant root loss, often resulting in tree death in 4 - 6 years. Trees designated for protection are worth saving and worth the cost for augering beneath the root system. Augering should be done in accordance with the enclosed specifications found in Figure 2 (3).

9. Tree watering. Provisions should be made in the original contract to water designated trees during periods of drought. Periods of drought are defined as a time when there is less than one inch of rainfall during any 30-day period. Trees should be watered at a rate of 50 gallons per inch diameter. Application should be made so that the water slowly soaks into the ground and does not run off.

Since most major construction projects are planned six months or more in advance of actual construction activities, trees that are to be preserved should be watered, fertilized, pruned, etc. during the fall or winter preceding construction.

10. Sidewalk removal procedures. Major injury to trees can occur when sidewalks are removed. Removal equipment may compact parkway soil, damage trunks and branches, and rip up roots. Broken sidewalk pieces should be pulled from the street surfaces with construction equipment, such as a gradeall, that is parked in the street, or in difficult areas, removed in wheelbarrows. Trees designated for protection should be root-pruned before sidewalk removal (see #6). When root-pruning, the on-site arborist should determine if severing roots close to the trunk on the sidewalk side will make the tree subject to windthrow. Any tree considered to be a windthrow hazard as a result of root pruning should be removed.

Damage to trees may occur during the sidewalk installation. Cement should be piped to the sidewalk and cement trucks should not be allowed to clean their chutes onto parkway soil. Valuable, healthy parkway trees can be protected by constructing the sidewalk around them. However, installing circular forms may greatly increase sidewalk construction costs.

11. Topsoil backfilling and grading. Existing excavated material is sometimes used for fill. Subsoil and fill material are generally unsuitable for root growth and should not be used in the root zones of parkway trees. Where topsoil is needed to finish grade, "good" topsoil, free of residual pesticides and containing low levels (less than 20%) of clay and 3-5% organic matter, should be specified. Parkways where trees are growing or will be planted should have a minimum depth of 12 in [30 cm] of "good" topsoil. In areas where "good" agricultural topsoil is not readily available, topsoil may need to be defined in stringent specifications. Patterson (4) has published a detailed definition of topsoil that can be used to develop specifications.

12. Mulch restoration. Tree roots do not grow well in competition with turf (sod) (1). When a municipality is making additional expenditures to protect trees from construction injury, homeowners should not insist on installing sod around these trees. Optimum root growth is obtained when the organic mulch is 4 in (10 cm) deep. The homeowner must be willing to commit time and effort to maintain the mulch (mostly weed removal). Barriers to contain the mulch can be used as long as they do not damage the lateral root system. For homeowners who are not willing to maintain the mulch collars, the areas around trees can be sodded or seeded with grass. The municipality should provide literature to residents warning of the consequences to the trees when using sod instead of mulch.

13. Post-construction pruning. Some trees may be inadvertently damaged during construction. Post-construction pruning removes any damaged branches or damaged trees not worth protection measures. Money for post-construction pruning should be part of the construction project.

14. Tree planting program. This is not a program to replace trees lost or injured because of construction activities. This is a program to fill every available planting site with a tree species that is specifically selected for each site. It should leave the parkway better than it was before construction.

15. Resident education program. Although not included in the original 14-point program for Glen Ellyn, this step is considered to be one of the most important. Each resident should receive information prior to construction which provides details of the tree protection program. Following construction, each resident should be given information on the proper care and maintenance of parkway trees. Residents should also be given a letter of appreciation for their patience and cooperation.

Conclusion

Because of Phase I construction damage, several trees along West Hill have been removed and several others are showing symptoms of decline and will need to be removed. The contractors were not held liable for the damaged trees because provisions were not made for tree protection in the planning stage.

In contrast, before the Phase II project began, most of the trees that were in poor condition were removed. After completion of the project, all available planting sites were planted with a suitable species. These trees will remain an asset to the residents for many years to come, and their appearance will not be altered by utility arborists. At this time existing trees show no evidence of injury or decline due to the construction project.

During Phase II some minor tree damage occurred and because the tree protection measures and fines were written into the contract, the Village was able to deduct the cost of remedial measures from the contractor's payout.

Infrastructure improvement projects, similar to this one, are common among municipalities throughout the country. This 15-point tree protection program should provide guidance for any municipality about to perform construction around parkway and public trees. It is especially useful for communities that lack the services of in-house professional arborists.

Literature Cited

1. Green, T. L. and G.W.Watson. 1989. Effects of turfgrass and mulch on the establishment and growth of bare-root sugar maples. J. Arboric. 15:268-272.
2. Matheny, N.P. and J. R. Clark. 1994. A Photographic Guide to the Evaluation of Hazard Trees in Urban Areas. 2nd Ed. ISA, Savoy, IL. 85p.
3. Morell, J. D. 1984. Parkway tree augering specifications. J. Arboric. 10:129-132.
4. Patterson, J.C. and C.J. Bates. 1993. Long-term light-weight aggregate performance as soil amendments. pp 149 - 156 In The Landscape below Ground. Watson & Neely (Eds). International Society of Arboriculture, P.O. Box GG, Savoy, Illinois 61874.
5. Watson, G. W. et al. 1995. Root Injury and Tree Health, Video. Illinois Arborist Association, 33 N. Waukegan Rd., Suite #1, Lake Bluff, IL 60044.

Construction Applications of Hydraulic Soil Excavation

Rob Gross

Hydraulic Soil Excavation is a method that can be applied to remove soil from near and among valued tree roots. Unlike most other soil excavation techniques, this process is gentle to roots and can preserve fine details. A carefully applied water stream can efficiently reduce soil into slurry, which may either flow away from the excavation by gravity, be removed by a trash pump or with a high volume vacuum truck. Appropriate water delivery for this process maximizes erosion rate and minimizes root damage. Hydraulic soil excavation has many applications before, during and after building and landscape construction. Exposed roots can be clearly viewed, accurately appraised, and treated as needed. This article covers the basics of hydraulic soil excavation, reviews many applications and details four specific, construction-related cases.

Construction activities near trees commonly result in negative impacts to trees roots and the soil environment, which may lead to premature demise of impacted trees. Because tree root systems vary dramatically from one site to the next and from one tree to the next, it is nearly impossible to accurately know where the roots are without first exposing them. Hydraulic soil excavation is an effective method of exposing roots for investigation and/or treatment. Roots excavated with this system exhibit less damage than manual, air or machine excavations (Fig. 1). To be effective and efficient, hydraulic soil excavation requires knowledge of soils, roots, hydraulics, and use of appropriate specialty equipment. Improper water application can result in a slow rate of excavation or avoidable root damage. Careful coordination and management of the many system variables for water delivery and slurry management in advance of, during and after the soil erosion process are essential.

Soil can be hydraulically excavated with an array of systems and system components. The excavation system described in this paper is detailed by Gross (5) and is specifically designed to be time efficient and minimize root damage when compared with other excavation methods.

Basic Components of Process

Excavation site preparation. To excavate as safely and rapidly as possible, remove impenetrable surface coverings and objects like bricks, cement, plastic and wire, as well as plants and other organic material from

---

Consulting Arborist, DenroTech, P.O. Box 766, Calistoga, CA 94515

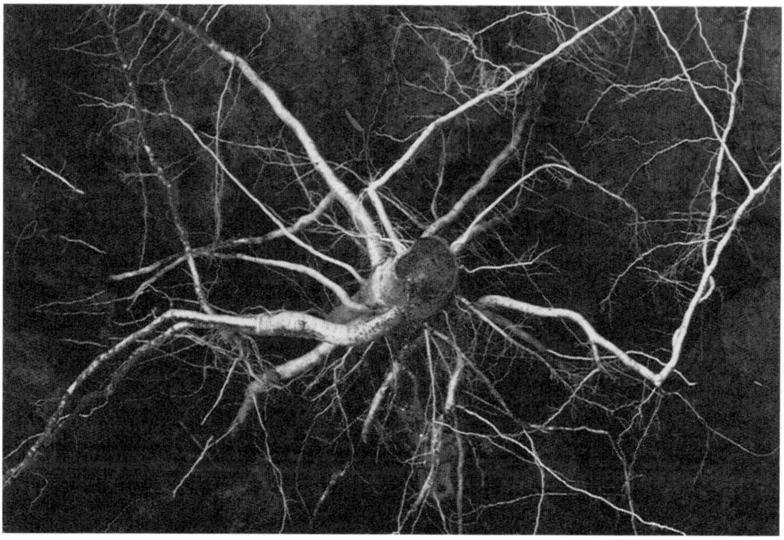

Figure 1. Hydraulically excavated root system of 4 inch Crape myrtle (*Lagerstroemia indica*). Note the fine detail on the small roots.

the soil surface. Pre-irrigate soil to saturation over the entire excavation area to maximize soil erodibility (pers. com. R. W. Harris). The entire excavation area should be saturated as deeply as the anticipated excavation. Fine textured soils, compacted soils and sites on an incline may require split-applications or multiple applications of irrigation water to attain adequate moisture penetration. Excavating dry soil is much slower because there is considerably more mechanical resistance to erosion. Also, when dry soil is excavated, water and slurry are strewn over a far larger area than if the soil were saturated. This commonly leads to secondary problems of increased root damage and increased potential of airborne projectiles which can harm property and personnel.

Water delivery system. The *water source* should have sufficient volume for the entire job. One excavator using this system may flow 10 to 20 thousand gallons of water each day. Sources vary and may be limited by volume, flow rate, pressure or access. Hydrants, lakes, streams, tanks or tankers can all be utilized; each has it unique advantages and constraints, such as the cycling time required for a tanker to refill. Some water sources are constrained financially by the provider who may require metering for compensation or environmental reasons.

Water pressurized 60 - 80 psi (pounds per square inch) at the nozzle, while flowing is the most desirable. Other delivery systems require different flows and pressures. It is important to carefully monitor pressures and know the flow rate it produces to maximize the rate of soil erosion and minimize root damage.

Water conveyance from the water source to the site utilizes a supply line 2.5 or 3 inches in diameter. The last fifty to one hundred feet should be reduced to a handline hose 1.5 to 1.75 inches diameter for maneuverability and to reduce friction loss and operator fatigue.

Nozzles vary considerably. Nozzle features should include 1.5 inch fittings, variable stream (straight to fog), bale type shut-off or "throttle," and pistol-grip type handle. Flow rate(s) of the water moving through the nozzle can be adjusted on select nozzles. The nozzle with manual flow rate adjustment (Fig. 2) is desirable for water delivery below the desired pressures and flow rates. This feature will allow the nozzleperson to manipulate flow to maximize rate of erosion. Automatic nozzles internally control the flow rate of water moving through the nozzle, which simplifies the operation. Fixed flow rate nozzles can severely limit rate of soil erosion and eliminate control of root damage. The twist type shut-off is both slow and unsafe. The pistol

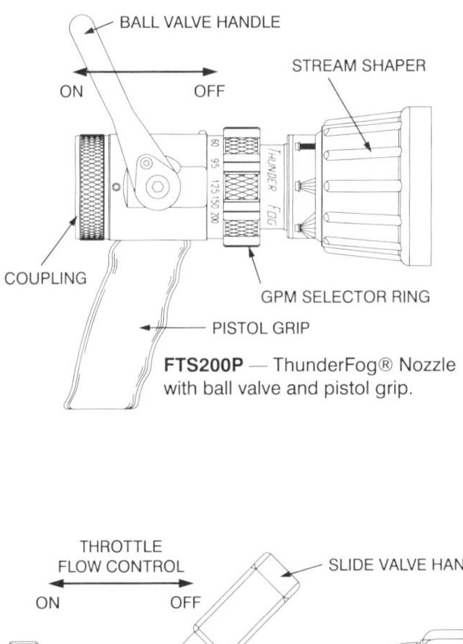

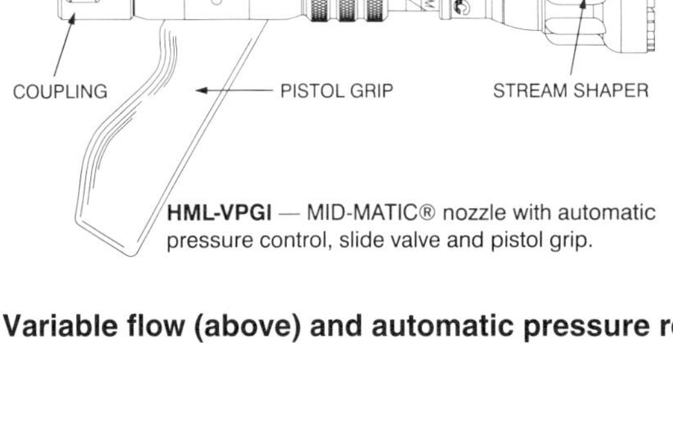

Figure 2. Variable flow (above) and automatic pressure regulation (below).

grip increases operator control and reduces operator fatigue. In situations where time is not a consideration, nozzle type is of less importance.

Slurry management. Once the soil is reduced to slurry by the pressurized water, it may require further considerations or management. On slopes, slurry may simply flow away, while on flat terrain or in low areas a pump or vacuum may be required. For small projects or when time is not of concern, slurry can be pumped away to a suitable disposal site. A high volume vacuum truck can enhance the rate of soil excavation. Slurry disposal is regulated in some jurisdictions, so this should be considered in advance.

Personal safety equipment. Measures always should be taken to protect: head, eyes, face, ears, feet, hands and body.

Applications Associated with Construction

- Locating roots for planning and construction can help avoid critical or sensitive roots or root areas. Root distribution is not symmetrical nor consistent, especially in urban soils, therefore this application is particularly helpful.
- Condition appraisals for diagnosis (3,5) of disorders lets the tree professional view underground conditions otherwise not visible.
- Forensic investigations can utilize this method to locate and appraise roots and root conditions as needed for legal cases. Damaged roots can be viewed clearly and documented once exposed.
- Hazard appraisal of disease impacts and structural flaws can be aided, as roots are exposed for accurate appraisal. This application can be particularly helpful for looking at the roots or root collars of a large number of trees, large sized trees or when detailed quality is required.
- Pre-climbing tree examination (2) calls for examination of trees before climbing to determine the amount of grade change, and to expose the root flare and lower trunk areas for structural flaws and disease involvement. Hydraulic soil excavation is appropriate for large trees and large numbers of trees.
- Research of roots, root systems or the root environment (1,6,8) can be accomplished with great detail, in a reasonable time frame (Fig. 3).
- Trenching (5) can excavate soil from between and below the roots. Commonly roots grow down to a specific and often uniform depth of less than three feet. Pipes and other conduits can be inserted below the roots exposed in the trench, or even threaded through the roots, and then the trench can be refilled. Hydrojet replaced soil can facilitate settling.
- Pier foundation installation (pers. com. R.W. Harris) can be accomplished by boring or "pot-holing" hydraulically. If large or numerous roots are encountered, the pier location can be easily and quickly shifted to a more desirable location.
- Sidewalk or foundation conflicts with roots can be accurately located, appraised and treated. When offending roots are encountered they can

Figure 3. Soil excavation using water delivered in fire hose on left with vacuum tube on right. The roots of these specimens were excavated 3 ft deep on 6 ft radius in about 40 minutes. This study looked at root response th three irrigation regimes at UC Davis.

be removed or managed. Root barriers also can be installed or appraised hydraulically.

- Compacted soil conditions can be improved by removing compacted soil layers or portions of the root area soil in trenches radiating from near the trunk out toward the dripline manually (9), using a trencher (4), or backhoe (9) to increase tree growth. Hydraulic soil excavation removes the soil while retaining the roots (5,10).
- Fill soil can be carefully removed in nearly any configuration or amount.
- Toxic soil removal (3,7) among valued tree roots can be accomplished without the root damage of conventional soil excavation techniques. To be efficient when removing large volumes of soil, special, high-volume water delivery and slurry removal systems are required.
- Root growth manipulation can be achieved by replacing soil in select areas or corridors for increased root utilization and enhanced growth. In situations where root damage is anticipated to create an environment that is well aerated, high in nutrients and has reduced physical impedance.
- Tree transplanting (pers. com. R.W. Harris) can be accomplished with root impacts minimized and the size of the roots ball can be reduced based on actual root system conditions. The exposed roots then can be cut carefully and boxed as needed. Large sized trees can be bare root transplanted during the dormant season without the weight of the soil.

Examples of Construction Applications

Case history #1. Hydraulic soil excavation was used in Davis, California on a Chinese hackberry (*Celtis sinensis*) to accurately locate roots in preparation for a new driveway installation. Once the soil was excavated and the roots were viewed, it was clear that the tree would be substantially damaged if the driveway were installed as planned. The driveway design and construction was altered as a result of the findings. In this case the excavation provided clarity on the actual location of the roots and allowed foundation piers to be located with minimal damage to the tree roots. With this method the findings can be so clear that it becomes difficult for decision makers to argue or disagree about what potential root impacts may result.

Case history #2. Two large diameter trees, one valley oak (*Quercus lobata*) and one coast live oak (*Q. agrifolia*), exhibited four large, severed root stubs near the base of the trunk. The owner was concerned about the impact the loss of the roots might have on these big trees. Careful observation suggested that more roots might also be damaged. Two hours of hydraulic soil excavation exposed extreme damage to almost 50% of the root area . Roots up to six inches diameter were damaged. The damaged roots had been torn away and were missing - only torn stubs remained. Presumably, this was caused by a tractor which was used to demolish the former structure. Careful observation of the damaged roots determined that two separate events damaged the root system, judging from the wound face color and the growth of cut roots. The damaged roots ends were cut clean, the soil was physically augmented and the soil water status was enhanced to improve root growth in to the impacted area.

Case history #3. A large, old and aesthetically critical valley oak *(Q. lobata)* is located by an large new home under construction. The root collar was not visible, indicating a raised grade, and the owner insisted on saving the tree at any cost. The general contractor was uncooperative and insisted that, "Nobody knows noth'n' about that tree." The root collar was excavated and a series of radial trenches were made using hydraulic soil excavation (Fig. 4). As a result, several root environment changes were discovered. The root collar had been covered with two feet of fill soil. Only part of the root area was at the original grade, while other parts were raised from one to over five feet with highly compacted fill soil. Two four-inch diameter PVC drain pipes had been installed, damaging more roots. A French drain brought water to the base of the trunk from the nearby turf area. After the hydraulic soil excavation, the root damage was cleaned, fresh soil was replaced in the trenches, and the French drain was re-routed away from the tree. When the contractor was shown the root conditions in clear view, he suddenly recalled many details about the fill, soil type, depth, degree of compaction and considerable other details of the impacts.

Case history #4. As part of a hydraulic soil excavation demonstration in Vail, Colorado, the roots of a lodgepole pine (*Pinus murrayana*) were ex-

Fig. 4. Soil impacts investigation with radial trenches. This process discovered numerous previously unknown impacts. The trenches were filled with fresh soil. This tree was over 200 ft from the vacuum truck.

cavated for clear observation and discussion. It was easy to observe that the roots of this pine were previously damaged by the installation of underground electric lines some time in the past. The roots were cut, scraped, torn and missing in some parts of the root area and not in others. Hydraulic soil excavation exposed a large wounded root at the base of the trunk and also a layer of one-half inch diameter roots which apparently grew after the soil was disturbed by the earlier construction. This excavation also created a potential utility trench where lines or pipes could be installed under the lower roots. The replacement soil was hydrojetted to facilitate soil settling.

Discussion

Hydraulic soil excavation is a powerful method of soil excavation with many applications before, during and after construction near valued trees. It can be efficient and cost effective and causes minimal root impacts when compared with other methods. Hydraulic soil excavation can be used wherever careful, gentle soil excavation is warranted. Soils of all types can be excavated in essentially any configuration or depth. For the process to be efficient and cost effective, both the water delivery and slurry removal systems must be carefully managed. When properly applied, hydraulic soil excavation provides a valuable tool for arborists, contractors, scientists, and anyone else working with or around valued trees and their roots.

Literature Cited

1. Baker, F. A.. 1993. Root disease reduces hydraulic conductivity of mature Douglas-fir and Grand fir roots, pp 166 - 175. **In** Proceedings of the 8th. International Conference on Root and Butt Rots of Trees.
2. Britton, J. C. et.al. 1995. Tree-pruning Guidelines. International Society of Arboriculture. Savoy, Illinois. pp 1 - 14.
3. Day, Steve. 1994. Hydraulic root excavation. Tree Leaves. Rocky Mt. Chapter, ISA. Summer. pp 4 - 8.
4. Day, S. D. and Nina L. Bassuk. 1994. A review of the effects of soil compaction and amelioration treatments on landscape trees. J. Arboric 10(2): 9 - 16.
5. Gross, Rob. 1993. Hydraulic soil excavation: Getting down to the roots. Arbor Age 13:10,12 - 13.
6. Lindsey, P., R. Gross and B. Milano. An investigation to assess the impact of street infrastructure improvements on the roots of adjacent cork oak trees. Submitted to Trees & Building Sites.
7. Miller, Randall H. 1994. Protecting contaminated trees. The Sharon Steel/Midvale Tailings Project. J. For. Oct.
8. Oltman, David. 1995. It's a blast. California Farmer. 278(10): 10,11, 24.
9. Watson, G.W. 1990. Tree growth revisited. Golf Course Management. 58(6): 7 -12.
10. Wootli, K. 1989. Radosan + Sanator. J. Arboric 15:253.

Ecosystems Approach to Construction Management

Steve Clark

A network of consultants and I have been pursuing a line of thinking that approaches the ecosystem with wonder. We wonder how to do a better job of building in concert with nature by considering the interrelationships that make up the urban ecosystem. Each of us must understand our responsibility in the ecosystem alteration process as we continue to build our cities. It is imperative to know how the ecosystem functions in order to do our part effectively and responsibly. We also need to understand, at least in principle, how the urbanized ecosystem functions effectively.

The cutting edge. Those of us in the ecosystem business are ready for some new challenges in the '90's. It is time for us to play a major role in planning and development in a nation that has been abused by greed, apathy, and ignorance. Education is going to take on a new dimension of graphics, style, and appeal that pulls people eagerly to truth, whether it is how to develop in concert with nature or how to fix the social ills of society.

I have worked in the field of urban forestry for 22 years, and in the process of conserving trees my colleagues and I collected a series of techniques that we changed and fine tuned along the way. I reached a point in tree conservation where I became bored. Then it finally hit me. I realized that there is no real integrity in getting better and better at something just to fill my bank account. The real cutting edge is helping someone else get good at his profession. .

I have learned that the strength of our team would increase when we joined other related fields. We are all dependent on each other to participate in quality stewardship of our nation. The latest quantum leap for me has been to be part of a network of consultants that views development from an approach that considers the carrying capacity of the ecosystem. We are totally involved from the beginning to end in the design and construction process. We develop our own drawings through an experienced AUTO-CAD operator who takes the drawing from preliminary stages to the final documents. One of our projects in Alabama had 15 sheets in the final document covering all aspects of the natural resource with detailed graphics on ecosystem conservation. Our team expertise included wetlands, geology, hydrology urban forestry, natural resources planning, landscape architecture, civil engineering, soils, and wildlife. We have been amazed to see that the cost of development goes down when you develop in concert with the ecosystem.

Steve Clark, Inc., Natural Resource Planning, P.O. Box 3425, Brentwood, TN 37024

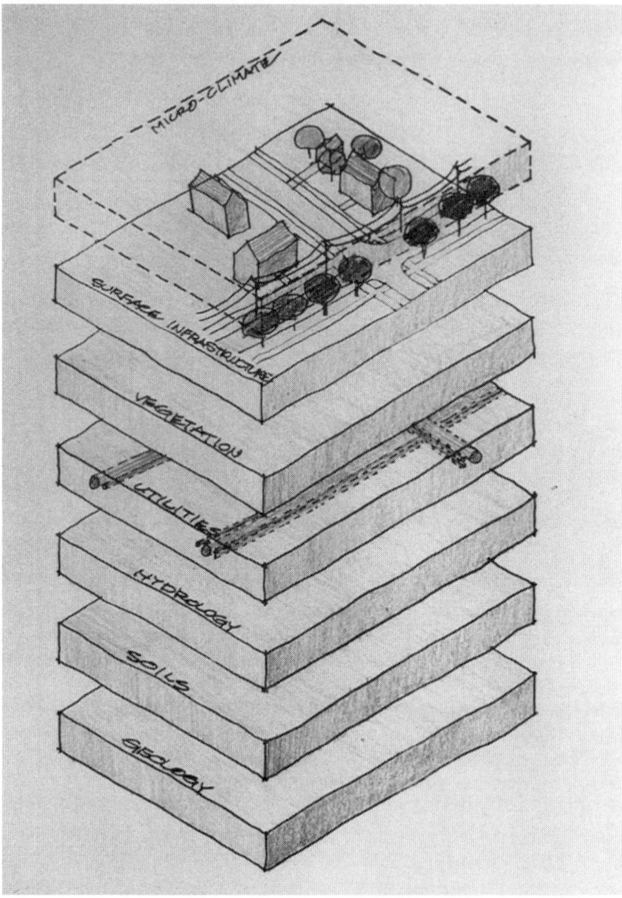

Figure 1. Ecosystem layers used to explain to laymen how the system works.

By using our ecosystem maps as a tool in the design process, we can guide the engineer in most cost, as well as engineering, decisions.

If we want to be on the cutting edge, we all need to constantly look for new territory, better techniques and ways to learn more, produce more quality, and help other professionals do a better job. You are in uncharted waters. You don't have all the answers and you are making decisions as you go. Our natural resource concept considers as many layers of the ecosystem as are appropriate for a particular site. The layers are generally geology, hydrology, soils, vegetation and micro-climate. Other layers, such as relief and surface drainage, may influence the negotiation with the architects, civil engineers, or other professionals.

Our team tries to avoid slipping into a specialty coma. In my experience most of us are over specialized. We don't hear or respond to other fields when we are in our specialty mode. Our team tries to carefully listen for the common ground in the overlap of professions. We have randomly estimated that it will take from two to four years for our team to actually walk on the common ground between.

Organized systematic approach. The 1970s created tremendous anger between the environmental activists (greenies) and the free enterprise developers and builders. The pendulum might not have swung so far if the environmental laws had not been written with an agenda based on half truth, no truth, and truth carried to the extreme. Even tree and landscape ordinances varied in their usefulness. When we show a united front of all the environmental disciplines, we then are ready to learn the detailed systematic approach that builders and developers already have. Recently, I made the statement to one of our newest clients that they had a good systematic approach to developing residential subdivisions except for one thing. They didn't have a systematic approach to ecosystems management that would give them a better product. This better product gives better quality of life and often gives greater profitability in the long run and sometimes in the short run.

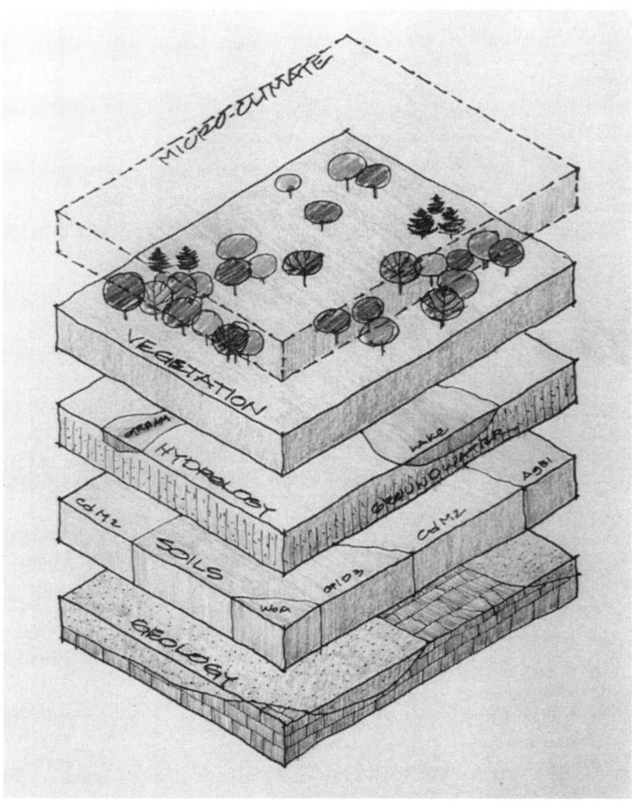

Fig. 2. Ecosystem layers with some of man's construction intervention.

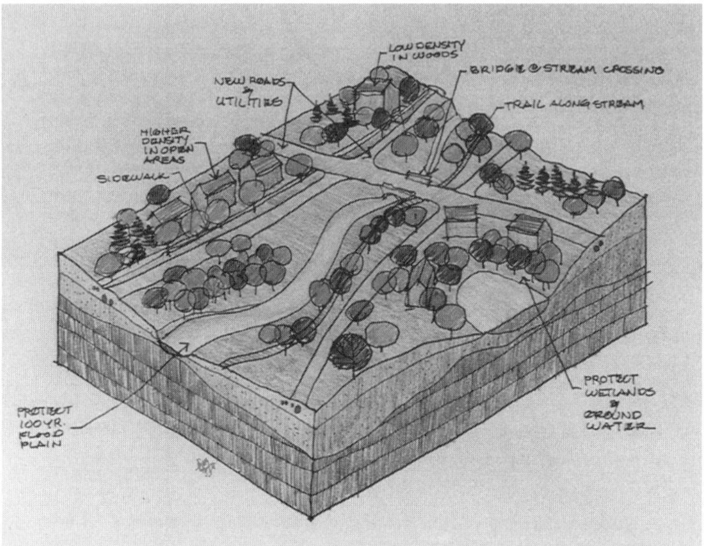

Figure 3. Concept of building in concert with the ecosystem.

With the right attitude we can endure the give and take negotiations, but without organization, unification, and plain "street smarts" we can't compete or contribute in a highly competitive business world.

If the tree is our only concern, we may win the battle, but lose the war. We must take a giant leap now to walk confidently with the developers of this nation, defending the ever so delicate ecosystem. It is not the time to be a fanatic. It is time to have a respected plan that works in the midst of free enterprise.

Are trees, more important than anything else? I think that urban forestry and related professionals are guilty of building our case based on half truth. We are so tree oriented that we are not telling but a portion of the story. Don't you think that the geologist, hydrologist, soil scientist, climatologist, civil engineer, landscape architect, and architect know the rest of the story? We must understand how we fit together. Why don't we take it on ourselves to start working on a systematic approach that is user friendly, cost effective, and ecosystem sensitive with a universal language that crosses all disciplines and mind sets.

I've been saying for years that trees are not the most important thing to the ecosystem. However, they do play a major role in producing a high quality of life. We've been mad for a couple or three generations at those who dominate and manipulate our environment, kill our trees, pollute our water and alter our forests until they are hardly recognizable. Much of the focus for many of us is driven by government programs and grants. Planting trees all across America is fine if they are in concert with the infrastructure and in a well thought out maintenance program that can be enforced. We need to take the facts and make logical decisions with stewardship as the overriding principle. We have wasted millions of dollars on environmental programs that were not well organized. Let's develop a system that saves the functionality of the ecosystem, yet doesn't kill the free enterprise spirit. It is time for everybody to decide if they want to be part of the problem or part of the solution.

We need to put true stewardship principles into tomorrow by handing over the ecosystem happy and healthy to the next generation. If we have learned from yesterday, take responsibility for today and develop a good stewardship program that gives people most of their conveniences systematically arranged and managed, then we are ready to hand the baton of good stewardship off to the next generation.

Communication, coordination and cooperation. As I travel all over

Figure 4. Three story condo built in concert with nature is designated as a wildlife sanctuary in cooperation with the National Institute of Wildlife and numerous other agencies.

this country, there are three words that describe the biggest needs—Communication, Coordination and Cooperation. On May 8, 1995, in Huntsville, Alabama, a 100 year old dogwood weighing approximately 77,000 pounds was moved with the most efficient communication, coordination and cooperation that I have witnessed in my entire career. The most exciting thing was the diversity of the players on the team. The U.S. Army, Alabama State Highway Department, county and city government with numerous departments, volunteers to furnish a crane, trucks and other equipment, schools, private enterprise and the media all participated as if they owned the tree and the whole operation was important to them. So it happened one time in history that a very diverse team consisting of public and private people rolled up their sleeves and communicated, coordinated, and cooperated to reach a common goal. Up until this time, I had been saying that people don't take time to communicate, coordinate, and cooperate. This project showed me that there may now be hope.

Even now, I think with all of our computers, videos, modems, faxes and cassette tapes, we need to study other professions to stay on the cutting edge. Communication is far more than talking. Money, power and authority are the major underlying motivation for coordination and cooperation in many positions in America today. Everybody has some motivation to coordinate with others. We must take responsibility for our team and others to look for ways to make the process easier for whomever we are coordinating with, whether it's planting a tree or building a skyscraper. Cooperation is the unconscious or conscious action of working together for a common goal. I must admit that when I'm in a tree preserve mode, I tune everything else out. I'm sure people think I'm semi-comatose when I answer all of their questions with—now the tree this and the tree that. Other professions have the same problem. The common ground between people is weak due to the lack of education about the other's profession. Education is the critical bridge necessary for whole hearted cooperation. We are obviously not going to teach other professions everything about our business, but if we let them walk comfortably in our camp and show them that we want to help them in their camp, we can build now with integrity and hand the spirit of cooperation to the next generation.

Abraham Lincoln said that people are first, the land is second and laws are third. If you ask the question on all projects, "what is best for the people and the land?" then your motivation is right. If the law is not blessing the people, or the land, then we need to either change it or do away with it. Cooperation is easy with the attitude of, "what can I do to help?" We must take responsibility for people inside and outside of our profession by looking for common ground to complement their efforts. We must enthusiastically look for synergism that promotes unity, esprit de corp, friendships and prosperity for all of us in a team effort. Any good sales person can sell America what it wants, but it takes integrity to sell it what it needs.

Conclusion

Working together is not a new concept and I certainly haven't brought any revolutionary thoughts to this paper, but I've just shared with you what I think we need to do to get involved in building and maintaining a nation. My message, I hope, is clear:

1. God has masterfully put this ecosystem together for us to use, but manage wisely for our generation, as well as the generations to come.
2. Your part is important, if you care about your work and you're in the right position.
3. We are all in the people business and we the people are in charge of the stewardship of this earth.
4. The land and resources are not more important than the people, however man will not survive if he does not manage properly the resources.
5. Laws are to protect the people and the land through an organized systematic approach that educates, educates, educates and then regulates only to keep a pattern of good stewardship for decent people that want to do the right thing. The last thing a law is to do is to stop the indecent people that don't care about the people or the land.

I want to end with a story. A boy was walking along a beach as the waves came in and went out and the boy started to run to pick up a star fish and throw it back into the ocean. A man came up to him and said, "What are you doing?" The boy said, "I'm trying to throw as many starfish back into the water as I can because when the tide goes out they will be stranded and die." The man said, "It doesn't matter. It is a fruitless effort." The boy as he picks up another star fish and throws it into the ocean looks at the man and with determination says, "It matters to that one." He picked up another and said, "It matters to that one." He picked up another and said, "It matters to that one." It matters to me. I think it matters to you.